RETHINKING DISPLACEMENT: ASIA PACIFIC PERSPECTIVES

Rethinking Displacement: Asia Pacific Perspectives

Edited by

RUCHIRA GANGULY-SCRASE
The Australian Catholic University, Australia

and

KUNTALA LAHIRI-DUTT
The Australian National University, Australia

ASHGATE

Published by
Ashgate Publishing Limited
Wey Court East
Union Road
Farnham
Surrey, GU9 7PT
England

Ashgate Publishing Company
110 Cherry Street
Suite 3-1
Burlington, VT 05401-3818
USA

www.ashgate.com

British Library Cataloguing in Publication Data
Rethinking Displacement: Asia Pacific Perspectives.
1. Pacific Area – Emigration and immigration. 2. Migration, Internal – Pacific Area. 3. Internally displaced persons – Pacific Area. 4. Environmental refugees – Pacific Area. I. Ganguly-Scrase, Ruchira. II. Lahiri-Dutt, Kuntala, 1956–
304.8'091823–dc23

Library of Congress Cataloging-in-Publication Data
Ganguly-Scrase, Ruchira.
Rethinking Displacement: Asia Pacific Perspectives / by Ruchira Ganguly-Scrase and Kuntala Lahiri-Dutt.
p. cm.
Includes bibliographical references and index.
1. Refugees – Asia. 2. Refugees – Pacific Area. I. Lahiri-Dutt, Kuntala, 1956–
II. Title.
HV640.4.A68G35 2012
362.87095–dc23 2012027618

ISBN 9781409453482 (hbk)
ISBN 9781409453499 (ebk-PDF)
ISBN 9781409484813 (ebk-ePUB)

MIX
Paper from responsible sources
FSC® C018575

Printed and bound in Great Britain by the MPG Books Group, UK

Contents

List of Figures and Tables

Figures

Tables

Notes on Contributors

Ruchira Ganguly-Scrase is currently the Professor of Anthropology and the Course Director for International Development Studies and Global Studies, Australian Catholic University, Melbourne. Till recently, she was the Coordinator of the Centre for Asia Pacific Social Transformation Studies (CAPSTRANS), University of Wollongong, Australia. With a PhD in Anthropology from the University of Melbourne, Ruchira has specialised on the intersections of forced migrations and various development narratives. Her book *Global Issues/ Local Contexts: The Rabi Das of West Bengal*, Orient Longman, New Delhi/ Sangam Books, London, 2001, explored the displacement and relocation of artisans in 19th century India and the subsequent identities formed in the process of resettlement. More recently her research has focused on the consequences of global neoliberal reforms for various communities and classes. She is the author of a number of articles and book chapters on migration, the impacts of neoliberal reforms, ethnographic method, childhood and schooling, and gender relations in Asia. Her most recent book is *Globalisation and the Middle Classes in India: The Social and Cultural Impact of Neoliberal Reforms*, Routledge, London/ New York, 2009 (co-authored with T. Scrase). Ruchira is also completing a major study funded by the Australian Research Council on the experiences of dislocation felt by garment workers in India and Bangladesh.

Kuntala Lahiri-Dutt is a Senior Fellow with the Resource Management in Asia Pacific Program at the Crawford School of Public Policy, The Australian National University. Trained as a human geographer, Kuntala has extensively researched and written on the social, environmental and gender impacts of the rapid expansion of coal mining, particularly the consequences of displacement of indigenous peoples in eastern Indian colliery areas and the displacement of peasantry. Her two books in this area are *Women Miners in Developing Countries: Pit Women and Others* (jointly edited with Martha Macintyre) published from Ashjgate in 2006, and *Gendering the Field: Towards Sustainable Livelihoods in Mining Communities*, ANU E Press, 2010. Her critiques of knowledge production in the discipline of geography, jointly with Saraswati Raju as the co-editor, were in *Doing Gender, Doing Geography: Emerging Research in India*, Routledge, 2010. Jointly with Gopa Samanta, her research on the river islands or charlands is being published as a co-authored book, *Dancing with the River: People and Lives on the Chars in South Asia* from the Agrarian Studies Series, Yale University Press in 2013. Another related field of investigation has been social justice and gender issues with regard to water management. These were published as *Water First: Issues and Challenges for Nations and Communities in South Asia* (jointly edited with Robert Wasson) Sage, 2008, and *Fluid Bonds: Views on Gender and Water* (edited), Stree, 2006. Kuntala teaches courses on Gender and Development, and Gender and Environment in Masters of Applied Anthropology and Participatory Development Program. Kuntala's current research interests include a rethinking of resource reliance and mineral dependence through the study of livelihoods in informal mining in developing countries.

Douglas Hill is currently a Senior Lecturer in Development Studies, Department of Geography at the University of Otago, New Zealand. Douglas was educated at the Australian National University and Curtin University and previously held an associate research fellowship at the University of Wollongong. He teaches and researches on the political economy of the countries of South Asia. His previously published work has examined issues related to water resource management, rural development, common property resources, politics and federalism in India, and the transformation of urban space and labour relations in the port sector in South and Southeast Asia. A book co-authored with Dr Adrian Athique examining the changing nature of urban space, leisure and capital in India's urban centres, *The Multiplex in India: A Cultural Economy of Urban Leisure*, (Routledge, London) was published in 2009.

Timothy J. Scrase is a Professor of Sociology and Associate Dean Research, Faculty of Arts and Sciences, Australian Catholic University in Sydney, Australia. Tim is a leading scholar on the impact of global social and cultural change in Asia. He has published widely on development and social change in a range of leading academic journals and edited collections, and has previously published five books. His most recent book is: *Globalization and the Middle Classes in India* (London: Routledge, 2009; co-authored with R. Ganguly-Scrase). Tim has been a visiting research fellow at the International Institute for Asian Studies (IIAS), University of Amsterdam.

Belinda Green is an independent researcher and educator in the area of health and social science. Belinda completed her doctorate in Social Anthropology in 2009 at the Centre for Asia Pacific Social Transformation Studies, University of Wollongong, Australia. Currently she works as the Health Educator and Refugee Advocacy training Officer at STARTTS, the NSW Service for the Treatment and Rehabilitation of Torture and Trauma Survivors and peak service for refugees and asylum seekers, Australia. She has extensive teaching and research experience at Universities of Sydney, New Castle, Macquarie and Wollongong. For the past decade she has also worked on a number of field based projects. In 2010 Belinda became the Co-Founder of *NSW Diversity Matters* —a Consultancy for Cross Cultural Awareness and Social Inclusion and continues her work in this area.

Pratyusha Basu is Associate Professor, Department of Geography, University of South Florida in the USA. Pratyusha obtained her PhD in 2003 on rural cooperatives and agrarian inequalities from the University of Iowa. Her latest book *Villages, Women, and the Success of Dairy Cooperatives in India: Making Place for Rural Development* was published in 2009 from Cambria Press in Amherst, NY. She has also published a number of journal articles, including in the *Annals of the Association of American Geographers* and *The Professional Geographer*. Pratyusha teaches Geography and also coordinates undergraduate coursework. Her current research interests include environmental politics, geography of cultural identities and border conflicts, international development, gender and feminism, rural-urban boundaries, agriculture and globalization in Asian countries, particularly in India.

Fadzilah Majid Cooke is Associate Professor School of Social Sciences, and the Ethnography and Development Research Unit, University of Malaysia Sabah, Kota Kinabalu, Malaysia. Fadzilah has worked in the area of agricultural and forestry politics and development as well as environmental change and customary land for close to 15 years since being awarded a doctorate in 1996 by Griffith University, Australia. She has published two books, made contributions to national as well as international journals in Europe, Australia and New Zealand and is currently leading the Sabah team for the Malaysian Human Rights Commission's (SUHAKAM) national inquiry into indigenous rights to customary lands.

Suneeti Rekhari is a Social Anthropologist and teaches at the Institute of Koorie Education, Deakin University, in Geelong, Australia. Trained in Sociology, Suneeti received her PhD from the University of New South Wales in 2006. Currently, she teaches Anthropology and Sociology courses, while researching gender and belonging from a postcolonial angle. She has published a number of articles on the representation of Indigenous identity in the past. Her most recent publication is an analysis of the Australian film 'Jedda', in Bennett, James E. and Beirne, Rebecca (eds), *Making Film and Television Histories: Australia and New Zealand*, 2011 (I.B Tauris).

Bokhtiar Ahmed is completing his Doctoral studies in Social Anthropology from the Faculty of Arts and Sciences, Australian Catholic University. After undertaking a Masters in Anthropology degree from the University of Dhaka, Bangladesh, and a degree in 'Master of Social Change and Development' from the University of Wollongong, he lectured on anthropology at the National University and the University of Rajshahi in Bangladesh. He has contributed to a number of national conference volumes and journals on issues ranging from anthropological paradigms in South Asia to language, identity and culture. His present research focus is on social change in South Asia in relation to identity, development and power.

Gopa Samanta is an Associate Professor of Geography at The University of Burdwan, India. Gopa received her Doctorate degree in 2002 on rural-urban interactions in eastern India, a research interest that she is still pursuing with financial assistance from the University Grants Commission, Ford Foundation and the National Agency of France. With funding from the Indian Council of Social Science Research and Australia-India Institute, Gopa has also researched women's empowerment through microfinance in eastern India. She is presently working on a book manuscript *Women, Money and Power: Lessons from West Bengal Villages*. She has been a visiting fellow in the French Institute of Pondicherry, India, in 2006 and in the Research School of Pacific and Asian Studies, The Australian National University, Australia in 2003-'04. She has an ongoing collaboration with Kuntala Lahiri-Dutt under which program they are exploring the various dimensions of gender, livelihoods and poverty in the *char*lands of river Damodar.

Roberta Julian is an Associate Professor and the founding Director of the Tasmanian Institute of Law Enforcement Studies (TILES) at the University of Tasmania. She holds a doctorate in Sociology. Her unpublished doctoral dissertation explored the changing meaning of ethnicity among post-war Dutch settlers in Tasmania. For over ten years she has researched the re-settlement of Hmong refugees from Laos. She has published widely in the areas of immigrant and refugee settlement, ethnicity and health, globalisation and diaspora, and the relationships between class, gender and ethnic identity. These research works have been published as book chapters and journal articles in *Race, Gender and Class*, *Asian and Pacific Migration Journal*, *Women's Studies International Forum* and the *Hmong Studies Journal*. She maintains a long-standing interest in the settlement of migrants and refugees in Tasmania and has recently completed a project, funded by the Australian Research Council, on community policing and refugee settlement that focused on refugees from Africa. She has a particular interest in forensic science and is currently the lead Chief Investigator in a 5 year ARC Linkage Grant with Victoria Police, the Australian Federal Police and the National Institute of Forensic Science that will examine the effectiveness of forensic science in the criminal justice system. Her most recent books are *Australian Youth: Social and Cultural Issues* (with Pamela Nilan and John Germov) (Pearson 2007) and the third edition of *Australian Sociology: A Changing Society* (with David Holmes and Kate Hughes) (Pearson 2011).

A.J (Tony) Simoes da Silva is currently the Associate Dean Undergraduate Studies and lectures in Transcultural Studies, Faculty of Arts, University of Wollongong, Australia. Tony obtained his PhD in English in 1997 at the University of Western Australia, Perth. He has authored *The Luxury of Nationalist Despair: The Fiction of George Lamming*. Amsterdam and Atlanta: Rodopi, 2000, and has co-edited (with Dennis Haskell and Ron Shapiro) *Interactions: Essays on Literature and Culture in the Asia-Pacific*, University of Western Australia Press, 2000. He is also an Academic consultant to the Terminological Dictionary of Postcolonial Literary Theory (U. de Aveiro, Portugal & U. of Roskilde, Norway); Advisory Board member of Partial Answers: A Journal of Literature and the History of Ideas (Israel); The Icfai University Journal of Commonwealth Literature (India). Dr da Silva is the author of numerous articles on post colonial literary theory, including his most recent paper 'Paper(less) Selves: The Refugee in Contemporary Culture', *Kunapipi*, 32 (1) 2008. 58-65.

Lynnaire Sheridan is at the School of Management, University of Wollongong in Australia. She holds a doctorate from the University of Western Sydney. She has lived in Australia, Argentina, Mexico and the United States. Her research in a Latin American context developed from a year-long student exchange to Argentina. She subsequently completed her Bachelor of Applied Science (Environmental Management and Tourism) before undertaking her Honours degree study; a comparative cultural analysis of visitor characteristics and behaviour in the national parks in Australia and Argentina. After working in the United States for The International Ecotourism Society, she undertook her doctoral studies on the unauthorized migration of Mexicans to the United States with the Latin American Studies Group, in the School of Management at the University of Western Sydney. This was subsequently published as a book *I know it's dangerous: Why Mexicans risk their lives to cross the border* with University of Arizona Press. Lynnaire's current research interests include investigating the socio-cultural facets of global migration, tourism and business management in order to identify key points in complex scenarios that may facilitate problem solving and promote positive social, ecological and economic outcomes.

Acknowledgements

First and foremost Ruchira Ganguly-Scrase acknowledges the financial support from the Australian Research Council's Asia Pacific Futures Research Network (APFRN) for an international workshop on refugees and displaced people in the Asia-Pacific region. A number of papers from the workshop were selected along with other invited papers for publication in this volume. We sincerely appreciate the reviewers who assiduously read the papers and gave valuable feedback. Both editors are also grateful to the then Director, Dr Tim Scrase of the Centre for Asia Pacific Social Transformation Studies (CAPSTRANS) at the University of Wollongong, for supporting and hosting the conference. This volume is the culmination of a number of years of field research, several grant programs and research assistants who contributed at different phases of the project. In particular we would like to acknowledge Dr Gillian Vogl, Trent Brown and Aditi Bhaduri. Both Trent and Gillian provided valuable insights at various stages of writing. We have been greatly inspired by the works of the distinguished scholar, Prof Ranabir Samaddar, Director of the Mahanirban Calcutta Research Group, an organisation dedicated to research, advocacy and activism on issues dealing with justice, human rights and forced migrations. We are indebted to David Williams for copy editing and Madhula Banerji for her technical editing. We also thank the anonymous referees for Ashgate for their valuable comments on the manuscript. Finally, we thank Katy Crossan and her team at Ashgate for their ongoing professionalism and patience for our project.

PART 1
Neoliberal Globalisation and Displacement: States Displacing the Poor and the Marginal

Chapter 1
Dispossession, Placelessness, Home and Belonging: An Outline of a Research Agenda

Ruchira Ganguly-Scrase and Kuntala Lahiri-Dutt

'*Rethinking Displacement*' is about the ways in which mobile subjects, shifting across borders and boundaries, experience their dislocation, recreate their homes and adjust and adapt to new places, and in the process construct new ways of thinking about places that give meaning to individuals and communities. By privileging the human experience of place and placelessness through grounded research, this volume contests the compartmentalised treatment in studies of displacement into exclusive categories such as 'development-induced' and 'internal' and 'external'. In the contemporary global order that is characterised by extensive and rapid movements of people, there is a need to explore the multitude of interconnected factors causing displacements that compel people to move, within their homelands or across borders. This book addresses this need by bringing together a critical examination of historical and contemporary accounts of the displaced through a variety of approaches ranging from ethnographic and qualitative research methods to document analysis and literary interpretations. We emphasise that although the forms and conditions of mobility are highly divergent, the lived experiences offer commonalities that encourage us to dwell upon and chart a new path in displacement studies. Each contribution to this volume adds new insights into the different configurations of displacement and placement, and offers fresh interpretations of forced migration and dislocation in today's rapidly changing world.

This collection of essays also demonstrates that everyday experiences of displacement and resettlement can potentially connect the various, often oppositional, strands that currently define this academic field. A number of researchers have challenged such categorisations, including researchers from the Global South and those who aim to engage with policymakers. Chimni (1998, 2009), for example, has consistently critiqued the imperialist nature of the 'Refugee Studies' project. He observes that the theory and conceptual framing of this interdisciplinary field has become the exclusive domain of scholars in the Global North, and that a sustained postcolonial or feminist critique has been conspicuous by its absence. This point is echoed in Scalettaris' (2007) call for attention to the inherent analytical weakness of defining people into fixed categories such as refugees, migrants, internally-displaced people, and so on. In order to carry out a rigorous analysis, this is a dangerous trap best avoided. Turton (2003, 2004) has reiterated that in trying to be 'relevant', the blind adoption by academics of the categories and concepts employed by policymakers is likely to be 'downright unhelpful'. The questions of culture and history, and the personal narratives of home and the everyday shaping of belonging must never be elided in framing migrants by simplistic discourses of documentation, status and integration (Binaisa 2011: 4). Even the World Bank analyst Cernea (1996) observes that the literature on refugees coexists with that on 'oustees' or on 'development caused involuntary displacement', but hardly ever communicates or mutually enriches each other: 'Concepts and propositions are not inter-linked, and empirical findings are rarely compared and integrated' (Cernea 1996: 294). For instance, notions such as global and local mobilities, or displacement/placement, have increasingly appeared as interdependent categories as contemporary theorists such as Nolin (2006) have attempted to transform concepts of locality, community, and nationality. Adding conceptual depth to this critique, Ahmed et al. (2003: 1) observe '[b]eing grounded is not necessarily about being fixed; being mobile is not necessarily about being detached'. Indeed, both place and placelessness come to assume quite different dimensions when seen through the lens of subjective experiences of place and the spatial practices of place-making. These experiences of connecting and reconnecting to places then opens up the space to contest the notion that mobility is ungrounded or that homes are unconnected to change, to relocation, and/or uprooting.

One of the arguments this volume puts forth is that the boundaries of poverty-induced migration, development-induced displacement, and forced mobilities often intersect and are, hence, blurred. This overlapping is particularly so in the context of late modernity when place and personhood do not seem to be bound up and where there exists a 'generalised sense of homelessness' (Said 1979: 18, quoted in Kibreab 1999: 385). The resultant topophilia, defined by Tuan (1974: 4) as 'the affective bond between people and place or setting', cuts across the categories, boundaries and artificial space-time divisions emblematic of the current age. Epistemologically, a move beyond fixed categories is possible through privileging, in one's analysis, the human experience of this affective bond between people and places. In this context, Cresswell and Merriman (2011: 5–10) undertake this task by producing critical analyses of practices, spaces and subjects of mobilities; their 'mobile explorations' focusing on the human subject and their experiences of place. Binaisa's analytical framework (2011), on the other hand, privileges origins – the interstitial points at the transnational social field; the circulations, dynamism and fluidity that flow through transnational practices across these social spaces; and lastly, the networks and ties to think through notions of 'home' and 'belonging' for the displaced. The epistemological shift is possible through the recognition of forced and economic migrations as closely related and often interchangeable expressions of global inequality and societal crisis (as observed by Castles 2003: 17). The complex processes of decolonisation and increased integration of the world economy have set in motion large-scale population movements that render meaningless distinct categories of dislocations. It is through the deconstruction of various bureaucratic categories that both the diversity and similarity of people's experiences can be exposed. These categories often render invisible gendered experiences of forced migration, which in turn have been compounded by neoliberal globalisation. Neoliberalism, with its focus on structural adjustment programmes, has resulted in reduced social spending, leading to impoverishment and eradication of social, welfare, and educational provisions to people in developing nations. Resistance to poverty cannot be separated from political resistance and persecution; thus, turning the political refugee into an economic migrant. In addition, millions of people are displaced each year as a result of various development programmes and policies.

Furthermore, the persistence of stable categories of the displaced as internal and external marginalises feminist analysis and gender politics. That gendered power relations are played out in place and is not revealed unless a diversity of disciplinary approaches and subject matters, seemingly unconnected and from widely separated parts of the world, are brought forth together. In attempting to highlight the fluidity and diversities of displacement experiences, this volume contests the universalisations of 'home' and belonging, 'migrants' and 'refugees'. A feminist reinterpretation also necessitates the rethinking of whose experiences are privileged and why. Indeed, women are underrepresented in refugee determination processes, in claims for asylum and in resettlement (Hyndman 2010). For this reason, there is an urgent need to explore women's shared experiences as refugees, as displaced and as mobile subjects. Feminist analyses, such as the one by Walton-Roberts and Pratt (2005), through testimonies based on extensive fieldwork, can unsettle the clubbing together of all transnational women into the simplistic category of 'economic migrants' and problematise the cultural politics of 'recovering' migrant experiences in quotidian ways. Moreover, to say that the experience of displacement demands a gendered analysis because the majority of displaced persons and refugees in the world are women and children, is only one side of the coin. Even in instances, as Bannerji (2010: 100–1) has shown, where men comprise those who actually *physically leave* one place for another, women redefine what the home comes to mean and what it means to belong to this contested site. This volume, thus, aims to also converse with the growing field of gendered politics of displacement.

The chapters in this book are, in most cases, based on ethnographic and empirical research carried out in countries across the Asia-Pacific region. They present multidisciplinary scholarship by bringing together diverse approaches, and engage in a dialogue with a number of separated 'fields' such as migration studies, refugee studies, transnationalism, and displacement studies. Paying particular attention to incorporating the voices of the dispossessed, they highlight the experiences of uprooting and resettlement of individual women and men. The chapters examine various forms of displacement that are taking place and uncover the causes underpinning forced movements of populations. They also demonstrate the difficulties associated with establishing the difference between environmental, economic, and political factors. They aim to show that the category 'environmental refugee' can obscure the very complex reasons underlying environmental disasters.

They highlight the relationship between forced migrations on the one hand and the chosen path of development followed by the state on the other.

One of the biggest challenges in analysing the experiences of the displaced is the ongoing tensions between highlighting the powerless status of victims, who are excluded and marginalised, and that of the valiant struggle underpinning contestation of their marginality. The contributors in this volume continually address these dualisms and lay emphasis on the agency of the displaced. While some chapters focus on aspects of suffering and injustice, others pay particular attention to the resilience and renewal of communities that were initially formed by forced relocation and victimhood. Therefore, the creative formation of diaspora and efforts to rebuild communities are a significant component of this volume.

Overview: Mobility as a Central Category in the Study of Displacement

In the contemporary global order, the systemic violence of modernity broadly and neoliberalism in particular against those who do not conform to their logics seems to be expanding. People are treated as infinitely moveable and uprootable, with displacement of people becoming an accepted by-product in the production of wealth and the creation of various forms of public order (Bauman 2004, Escobar 2003, Siu 2007). As this tendency gains pace and begins to command more attention from researchers, Malkki's (1995) insight is pertinent in that traditional analytic categories of 'refugee studies' are not sufficient to recognise the complex nature of the situations that give rise to displacement, the differences between them, and the differential effects they have. Significantly, it has been noted that the framing of mobile populations who flee violent situations as archetypal victims not only serves to obscure the agency of the displaced, but completely neglects the position of those who have been unable to flee – people whose mobility has been constrained (Lubkemann 2008, Malkki 2005). When we consider this in connection with the fact that the mobility of those who have been displaced is constrained in various ways – from experiences of discrimination and persecution by local populations to the rules and regulations imposed by nation states – movement becomes a key concept in analysing the situation of those who have been subject to this form of systemic violence. In considering factors such as the differential impacts of displacement on men and women, various factors lead to the confinement of displaced women to the domestic sphere, providing them with less access to employment and services through which they could improve their condition (Daley 1991, Meertens and Segura Escobar 1996, Manchanda 2004). The field of border studies conceptualises movement as involving both mobilities and entrapments. In this regard Cunningham and Heyman (2004) provide a number of important theoretical tools, which prove highly useful in understanding the experiences and opportunities available for those subject to displacement.

Lubkemann's (2008) consideration of 'involuntary immobility' provides an important starting point for this perspective. He begins with the observation that most approaches to the study of displacement have brought only mobile populations into their analysis (2008: 456):

> [A]n entire interdisciplinary subfield (most often termed 'refugee studies' or 'forced migration studies') has coalesced and developed around the largely unquestioned propositions that migration is a requisite aspect of 'displacement' (i.e. to be displaced one must always have moved) and that displacement is inevitably the product of movement under crisis conditions (i.e. if you migrate in a crisis context you will suffer displacement).

Lubkemann delinks migration from displacement in order to better understand their relationship. Drawing on the case of the Mozambique war, which resulted in significant displacement, he shows how forced migration did not always have the anticipated effects of meaning exclusively disempowerment and loss. Those who did not leave were confined to their villages by both government and insurgent forces. This not only had the obvious adverse consequence of exposing them to war, but the constrained mobility prevented access to agricultural lands. In the context of an intense drought, to rely solely on local food production made survival precarious. On the other hand, while female migrant populations were found to be adversely affected by being physically displaced, male migrants, many of whom had extensive prior experience as migrant labourers, did not find the process as unsettling, and indeed, found many new opportunities in their forced relocation – opportunities not typically associated with 'displacement', which is taken to mean only disempowerment and loss. Drawing on this example, Lubkermann points to the importance of investigating prior experiences of mobility in understanding impacts on migrants.

Modernity, Neoliberalism, and Displacement

Official discourses justifying displacement are often framed in the name of development. This raises some unique issues. Oliver-Smith (2009) highlights some of the unique features of development-induced displacement. Unlike the effects of war or natural disasters, development is not officially recognised in negative terms. Indeed, development is seen as something beneficial, being 'in the national interest', rendering invisibility to the sufferings of the displaced people. Furthermore, there is usually no possibility of returning home after such kind of a displacement. Development, as a cornerstone of the modernist project, raises key questions about the relationship between modernity, development, and displacement. This relationship has been explored from a number of angles, as is considered ahead.

In a review of literature on development and displacement from the 1990s and early 2000s, Dwivedi (2002) identifies two distinct schools of thought. The first, which he terms the 'reformist-managerial' school, views development as necessary, and displacement as an unintended but inevitable outcome of the development process. The focus of research here is on the consequences of displacement, with the intention of developing procedural means by which the negative consequences of displacement can be removed or ameliorated. For the most part, this approach is favoured by planners and managers. This approach has been associated with the work of Cernea (2003). Dwivedi (2002) refers to the second school of thought on displacement as the 'radical-movementist' approach. This perspective, which often informs social movements against large development projects, views the phenomenon of displacement as a symptom of a development model in crisis and is the outcome of 'structural biases' within the dominant development paradigm.

On the 'reformist' end of the spectrum, Cernea (2003) has been developing a model for appropriate resettlement of the displaced that acknowledges the challenges faced by people in re-establishing themselves economically. He views displacement as a 'pathology of induced development' (2003: 37), but also recognises the indispensible nature of projects which produce great public benefits, and that these will inevitably, in some instances, involve the displacement of people. Further, he assumes that the numbers of such projects can only increase with increased populations and urbanisation. What he finds unacceptable is the fact that the situation most people face in the post-displacement context is impoverishment. One cannot justify projects in the name of poverty reduction, if they simultaneously give rise to poverty for others. Looking at the reasons why the overwhelming majority of displaced people end up in a worse condition than they were in prior to their displacement, Cernea finds that the principle of monetary compensation for losses in assets has proven to be grossly inadequate. It is not only the case that corruption and administrative flaws prevent people from receiving the compensation that is due to them, but also that a monetary payment cannot be considered equivalent to the loss in security of employment and of opportunities for growth that may have occurred had the displacement not taken place. As such, Cernea's alternative model places emphasis on the need to invest in the accelerated development of displaced communities, such that they have appropriate livelihood opportunities and are not left behind the wider economy in the time it takes for them to resettle.

Unlike Cernea, many scholars see development-induced displacement as not a merely accidental feature of 'development gone wrong', but as an inherent story of modernity in general and of neoliberalism in particular. This is explicitly discussed by Zygmunt Bauman in his book, *Wasted Lives* (2004). In this book, Bauman argues that the production of wasted humans occurs through displacement, an inevitable by-product of modernity. Thus even before today's globalisation, displacement has been a part of the project of colonial modernity. In its endeavour to create universal, rational order, modernity inevitably excludes those who are, for whatever reason, unable to conform to any designated place within that order. There will always be some people who cannot fit within the bureaucratic categories designed for the creation of social harmony. The issue is compounded by the modernist commitment to 'economic progress' (read, perpetual growth), which in its drive for ever greater efficiency renders traditional livelihoods redundant, uncompetitive, and systematically devalued. In many instances, displacement is intrinsic to development itself. For example, in postcolonial India, the state pursued a dualistic development agenda comprising large-scale manufacturing and agricultural modernization together with the promotion of small-scale enterprises (Jha 2001). The initial strategy, influenced heavily by the Gandhian legacy of rural empowerment, was to be achieved through the creation of village committees or panchayats (Gupta 1989). Yet the valorisation of small-scale enterprises was to prove merely symbolic devoid of concrete programmes to achieve empowerment of millions of artisans as intended originally in the Gandhian framework. Ultimately, those who benefited were large-scale manufacturers in the modern sector, displacing those in traditional occupations. The outcasts of modernity and development are those dispelled from their traditional life-worlds. However, occupational displacement is an important area that is often overlooked in studies of forced relocations.[1]

Rarely are the struggles of artisans to maintain a steady income from the production of traditional artisan goods satisfactorily met with by policies that ensure maintenance of their cultural heritage and identity through such crafts. Although governments in developing countries may attempt to assist small-scale enterprises, the structure behind the strategies implemented is usually insufficient in that artisans are unable to compete with large factories in the formal sector. Several case studies from countries such as Bolivia (Eversole 2006; Wethey 2005), India (Ganguly-Scrase 2013; Knorringa 1999), Mexico (Cohen 1998; Cohen and Browning 2007), Ecuador (Korovkin 1999), and Costa Rica (Wherry 2006) demonstrate the necessity for appropriate support for traditional craft producers in the form of effective government policy. Without the implementation of structured, efficient strategies, artisans face the dangers of displacement becoming impoverished in both their economic livelihoods and cultural practices.

Bauman (2004) emphasises that the twin processes of creating a universal rational order and of economic progress are amplified within the context of neoliberal globalisation. In reducing the restraints on transnational corporations, states have lost their power to control and regulate their displacement-causing effects. The commitment to perpetual expansion of opportunities for profits has meant that workers have had to accept a position of permanent precariousness (a condition which Bauman had previously referred to as 'liquid modernity'), giving rise to a greater number and diversity of 'economic refugees'. States have become powerless to prevent displacement from occurring, and are left simply with the task of managing the flow of people that it causes. Bauman notes the irony that displaced people from developing countries, who are forced to seek asylum in more developed countries, often become targets of those suffering from the same form of precariousness. Having been associated with transnational criminals, refugees are also treated as effigies for crime phenomena, which are completely beyond the capacity of states to control. With reference to the contemporary

1 See Scrase, this volume.

condition, Bauman emphasises that while displacement has always been a part of modernity, neoliberal globalisation has made the issues it generates more serious and inescapable. Previously, modernist countries were able to push their outcasts into other parts of the world. Now, there are no longer any such 'dumping grounds'. There may be physical space into which people may be pushed, but there is no social 'outside' anymore, as modernity has become ubiquitous. Consequently, finding the means to manage the 'human excess' produced by modernity has become an increasingly prominent feature of governance.

Evidently, much of the development-induced displacement today is characterised by the changing nature of the state. Over the past two decades globalising economies have been characterised by market liberalisation in the form of free flow of goods and capital. Subsequently, nation states have implemented policies of privatisation and deregulation. In this volume we will address this theme in number of chapters.

However, one should be careful in treating neoliberalism's grasp as a homogenous reality, experienced in the same way in diverse contexts on the one hand, and on the other unquestioning acceptance of the 'powerlessness' of the modern state. Here, the respective insights of Kingfisher and Maskovsky (2008) and Randeria (2003) assume critical significance. The former suggest that in some anthropological studies, neoliberalism has become a 'buzzword', used uncritically to describe an abstract and totalizing entity which imposes itself on populations in a relatively unproblematic way. Kingfisher and Maskovsky (2008: 188) highlight the need to see neoliberalism instead as an ongoing process, and as one subject to various forms of limitation and resistance in being applied locally. They stress the fact that while neoliberalism may have 'totalising desires', these are never realized because the social, cultural, and economic terrains are always heterogeneous. Focusing on factors such as subaltern resistance, limits to the process of subjectivisation, and the irreducibility of existing institutions to neoliberal logics helps to situate and problematise the neoliberal project. Randeria (2003) notes in the context of neoliberalism, the state (particularly in the developing world) behaves as a 'cunning state', using its perceived powerlessness with respect to international institutions on the one hand, and pressure from its own constituents on the other, as a way of behaving cynically and acting in its own interests.

A major source of displacement in the contemporary world order is large-scale development projects. Gellert and Lynch (2003) outline the breadth of displacing effects that such 'mega-projects' can have. They emphasise that displacement should be seen not only in terms of those who are immediately displaced by the imposition of the project itself. Large projects also have major *physical* displacing effects, disrupting soil, hydrological flows, and biodiversity. These physical displacements may have considerable impacts on people's lives and livelihoods, giving rise to *secondary human displacements*. Since the physical effects of mega-projects are so difficult to determine in advance, secondary human displacements are somewhat unpredictable, and, thus, more difficult to manage. This situation is not helped by the fact that, as Gellert and Lynch note, those who are most influential within decision-making organizations and groups are often informed by reductive modernist ideologies, which render invisibility to these secondary physical and human effects of development.

Fernandes (2007) has further explored some of the effects of these mega-development projects, with reference to several projects in India, but in particular the district of Singur, which had been the proposed site for the Tata Nano factory.[2] The evidence he has collected demonstrates that the number of people displaced by such large projects often greatly outweighs the number of employment opportunities created and that the jobs that are created are often inaccessible to the displaced themselves as they require technical skills (with training opportunities for the displaced rarely being offered). Looking at the situation of those who have been displaced, Fernandes finds that the average income post-displacement is approximately half that of pre-displacement, that people often have to resort to desperate and unsatisfying forms of employment, and that compensation (let alone rehabilitation) often never comes, or does not come in the appropriate form. Considering these factors the fact that the project at Singur was to take place on productive agricultural land, which, under previous legislation, could not be appropriated, Fernandes argues that the sole criterion for whether or not these mega-development projects are to go ahead has become their capacity to generate private wealth.

This example shows that, at least at a social level, the benefits of development-induced displacement are being outweighed by the costs, a point which a number of authors have explored. Escobar (2003) asserts that the gap between the system's need to displace people and its capacity to manage the effects of displacement, not just in terms of rehabilitation, but also to the provision of infrastructure and services to manage the movement of such great numbers of people, is increasing. The growth in this gap, he argues, is a function of neoliberalism, and its need for more intensive forms of accumulation. However, although he sees the gap increasing, Escobar does not see the problem of displacement itself as being particularly recent; rather, he stresses that it is an *inherent* aspect of European capitalist modernity, and not a merely incidental managerial failure. Capitalist modernity depends on the capacity to uproot populations in the endeavour to gain control over resources and respond efficiently to market demands. In order to achieve this, it is necessary that even in cases when people are not *physically* displaced, displacement still occurs *culturally*, by severing people's cultural connections to place and community, ensuring that they remain in a condition of being always potentially moveable.

---

2 Following extensive agitations, however, Ratan Tata eventually withdrew his proposal for construction at Singur, instead turning to Gujarat. See http://news.bbc.co.uk/2/hi/7657005.stm

Representation

Dwivedi (2002) suggests that the division within the literature has had considerable consequences in the way in which issues are framed and goals conceptualised. Where the 'reformist-managerial' camp does not investigate or question the assumptions of the dominant development paradigm, the 'radical-movementist' camp is openly critical of it, and advocates alternative development paradigms. Where the reformists are optimistic about the possibility of good rehabilitation outcomes, radicals typically view adequate rehabilitation as impossible. This openly critical stance often means that those taking a more radical stance are kept at a distance from policy and decision-making circles. Furthermore, Dwivedi suggests that when used as a means for informing policy, the 'reformist-managerial' approach impedes adequate conceptualisation of the holistic nature of the problem. Specifically, Cernea's top-down managerial approach to conceptualising 'risk' leads to a simplification of the experience of displaced people. Projected risks relate to loss of assets and sources of income without due sensitivity to more complex experiences of deprivation, such as lack of autonomy within the process and the loss of social bonds and traditional homelands, which are, strictly speaking, invaluable. This issue is compounded when, through lack of participation of local people, those who make the decision to take a risk are in a radically different position to those who bear the risk (a distinction made by the World Commission on Dams).

The consequence of this division in knowledge can be quite serious for displaced people themselves. This can be seen in Amita Baviskar's (2005) book *In the Belly of the River*. She explores the way in which the issues affecting those displaced by the Sardar Sarovar Dam project were articulated by the state and by activists within the Naramada Bachao Andolan [Save the Narmada River Movement], respectively. The state emphasised the necessity of development, the duty of the local people to accept the 'sacrifice' they would be making towards development and the official authorities that they would have to appeal to in order to achieve compensation. In contrast, the Andolan insisted that compensation would not be received or would not be adequate and that the project itself was based upon a flawed model of development, and, therefore, should be opposed outright. The people of the Narmada valley, caught between these two discourses, ultimately had no recourse but only to compensation, with the Andolan becoming a lengthy case in the Supreme Court to oppose the construction of the dam outright, rather than providing them with a voice in their struggle for rehabilitation.

Trade unions have often provided poor or contradictory representation for displaced people. In a study on migrant workers, Trimikliniotis et al. (2008) observe that trade unions in South Africa often express sympathy for migrant populations, while in practice support policies that restrict their open movement. Explaining this apparent contradiction, they suggest that unions saw the state and employers using migrants as a tool to divide the union movement, and to prevent this divisiveness, publically expressed their support for migrant labour.

The problem is that nationalised unions view the employment of migrant labour as a threat to the jobs of citizens, driving down available jobs, wages, and working conditions. As such, unions have supported policies that restrict migration and prevent employers from hiring illegal migrants – in stark opposition to the interests of the migrant populations themselves.

Nation States, Borders, and Movement

Despite the hype of globalisation and claims of a 'world without borders', nation-states remain important structures. Indeed, they provide the central context through which the claims of displaced and potentially displaced people are negotiated. Ghosh (2006) highlights the importance of recognising histories of struggle within nation-states against the recent uncritical embrace of de-territorialised discourses of human rights coming from many seemingly progressive quarters. His study examines the struggle of *adivasi*s (tribal people) in India against displacement due to big dams. He stresses that the discourses of resistance to development-induced displacement by adivasis have evolved in close connection to the discourses of colonial and post-colonial states in relation to 'tribals'. The British colonial state treated adivasis as outsiders, who were subject to exploitation from the mainstream Hindu community. Thus, they needed certain state protections – access to designated land, forest, and so on – which in turn reinforced their segregation. In their struggles against displacement, adivasis have drawn on their historical status as 'outsiders', in order to assert their unique needs. Ghosh also notes how, in their struggle for resettlement and compensation, adivasi leaders have drawn selectively on the common stereotype that 'adivasis can't handle their money' in order to make claims for land-for-land (rather than cash) compensation. He argues that in contrast to these forms of struggle which have developed organically through history, transnational discourses of 'indigeneity' tend to overlook these nationally and regionally specific histories and impose a kind of subjectivity that involves exclusions, which, in the case of Jharkhand, have proven to be harmful to the struggle against displacement. Focusing on the category of 'indigenous people', who are treated as a singular entity throughout the world, is to assert the unique formal–legal rights of the 'indigenous'. Such rights must be negotiated through the courts, which only middle-class, educated adivasis are able to access, to the exclusion of all others. Furthermore, transnational discourses of indigeneity place emphasis on the origins and continuous identity of the indigenous as grounds for claiming rights to land, access, and so on, in contrast to the historically embedded struggles of tribal people against the state-placed emphasis on ongoing experiences of exploitation and exclusion from both the state and the mainstream community.

A number of authors have recently considered that the transnational movement of people is mediated through *borders*. Cunningham (2004) sets her analysis on this topic against the 'iconography' of globalisation that proliferated in the 1990s, in which the 'new', 'globalised' world was seen as one characterised by free movement, mixing and fusing of cultures, hybrid identities, and so on. In the wake of the terrorist attacks in the USA on 11 September 2001, this public image of global society was abruptly revised. The idea of a leisurely freedom of movement was qualified by the need for 'national security', leading to a resurgence of the notion of borders as an important political category. This shift in public perception brought to light the fact that the movement that seemed to define global society was in fact a highly regulated kind of movement and one which could only be enjoyed freely by a very small minority, particularly in the traversing of international borders. Building on this insight, Cunningham effectively develops a critical perspective on how power operates on the 'flow' of people, goods, capital, etc., at the site of borders. Drawing on her own research on the movement of people at the US–Mexico border, she reveals the impact of policies and popular discourses on the possibility of movement, as well as the lived experiences of people.

Cunningham (2004: 346) goes on to argue that while the regulation of flow at borders is to some extent a part of the self-interested exertion of power on the part of nation-states, it should also be seen as part of the power structure of global society more broadly:

> [B]orders must also be regarded as a fundamental part of what might be called the machinery of mobility and enclosure in a global landscape and they function not in isolation, but in coordination with larger systems that secure capital and wealth in our global village.

Her views are further complicated by bringing in the effect of social movement actors, who incorporate the perspectives of local communities and/or human rights discourses in their endeavours to change the mechanisms by which flow is controlled.

Cunningham and Heyman (2004) weigh into this debate by conceptualising borders as sites of movement. They consider movement to involve not only mobility, but also its inverse – 'enclosure' as the restriction of movement. At the site of the border, rules are formulated, enacted, and negotiated concerning who has the right to mobility and who does not. This highlights the fact that although borders and nation-states are social constructs, which in some ways are becoming less relevant with globalisation (freer flows of commodities and capital, etc.), they continue to exert a powerful influence on people's lives. They also point out that although, as much previous research has identified, borders are often meeting points of diverse cultures and identities, they are also sites at which the state enforces difference. At borders, the bureaucratic categories for the regulation of movement are reified and differential identities are constituted. However, this is not a one-way process, and these locations can also be sites at which identities are contested.

Although their particular focus is international borders, Cunningham and Heyman (2004) make note of the fact that these processes do not occur *exclusively* at borders. Indeed, there are numerous cases in the literature in which similar instances of mobility and enclosure can be identified occurring within national boundaries, and also cutting across them. Pallitto and Heyman (2008) outline how, particularly since 11 September, 2001, the US has developed a complex network of checkpoints to control the flow of movement within its own borders. Border-like security practices, such as documentation, the use of identification technologies, screening and so on can be observed across the transport industry – whether on planes or highways – within the US borders. Pallitto and Heyman (2008) argue that this extension of movement control to domains other than national borders has meant that factors other than citizenship status come to determine the approval for or denial of entry into countries. New technologies of identification and surveillance have been developed and are applied at borders, airports as well as in more distributed networks. The authors highlight the fact that people end up with different *rights* (to cross borders, checkpoints), different *speed* of movement (whether or not one is held up by various screening procedures), and differential exposure to *risk* (including the risk of being subject to prosecution). Inevitably, this differential treatment will amplify the perception of social differences, having flow-on effects to people's modes of identification and behaviour.

Control over movement within state boundaries is not exercised exclusively by nation-states; often, the activities of non-state actors serve to question state sovereignty. Landau and Monson (2008) explore several ways in which the activities of both migrant and local populations challenge and reconfigure state-sanctioned principles governing the movement of people and rights to access territory. Studying the case of Johannesburg, they show how the xenophobia of sections of the local population towards migrants living in settlements coercively generated movements, which were in contradiction with the state's official commitment to inclusion. Local citizens, who perceived the migrant population to be the source of a wide range of social ills, took to vigilantism, using violence and the threat of violence to encourage migrants to leave the location. The actions of these citizens served to legitimise discriminatory actions by officials in the provision of services such as health and education. Thus, despite state commitments to provide such services for migrant groups, the actions of local citizens served to prevent their inclusion in the community. This form of enclosure was then amplified by the actions of the migrant communities themselves. In response to the violence committed against them, they drew on discourses that reified their differences from the mainstream community. Drawing on pan-African identities or their perceived identity as refugees (an identity that was taken on board irrespective of whether or not official refugee status was granted), many were able to justify their refusal to integrate into the South African community, reaffirming their physical segregation. In this way, one sees various forms of non-state agency defining possibilities of mobility and enclosure.

Some studies, however, reveal that this separation of citizens and states is not so clear-cut, and that often these kinds of actions on the part of citizens, though contrary to official state doctrine, may still accord with state interests. Also considering the situation faced by migrants in South Africa, Trimikliniotis et al. (2008) note that while post-apartheid South Africa remains in principle committed to multiculturalism, in the arena of immigration policy, it retains many features of the apartheid state. The difference is that where the apartheid state retained tight control over immigration in the interest of retaining racial domination, the new (neoliberal) state does so as a means of exerting control over labour. The migration of impoverished people from other countries in southern Africa is of benefit to capital, providing a source of cheap labour; however, the preference of capital is that this migrant labour be *illegal*, not only because it keeps wages down, but also because it precludes access to government protection from exploitation. Trimikliniotis et al. (2008) cite cases in which the state, in service of the interests of capital, has selectively presented migrant labour in such a way that contributes to their persecution by the general public.

Does Gender Matter?

The differential effects of displacement on men and women have only been considered relatively recently. Some early studies in this field suggested that men were more adversely affected by displacement than women. For men, displacement meant a loss of livelihood and position as provider and primary decision-maker within the family, whereas women's roles within the household were said to be less severely affected by relocation (see Hitchcox 1990). Colson (2003) has argued that this position neglects the numerous cases in which women define themselves through their roles in activities outside the domestic sphere, and that changing gender roles can lead to conflict within the household, which has a more adverse impact on women (see, Colson 1999, Huseby-Darvas 1994).

Legislation governing the treatment of the displaced often excludes the perspective of women. As Ganguly-Thukral (1996) outlines, policy in India recognises the family as the smallest unit of analysis when considering those affected by displacement, with the eldest male being the recipient of compensation. Policies issues by state governments for the allocation of compensation for particular instances of development-induced displacement often make no mention of women. This serves to reinforce women's lack of control over resources. In India, by law and custom, women receive no recognition as land owners; thus, following displacement, they cannot be recipients of land compensation. This is particularly problematic when, as is not infrequently the case, women are effectively the primary managers of family farms. Their displacement thereby often implies losing control over their main economic activity.

The construction of the Tehri Dam in Uttarakhand, India, resulted in displacements that had very unequal effects on men and women. Based on an ethnographic study of several rural villages in the region both pre- and post-displacement, Bisht (2009) argues that women experienced displacement in a qualitatively different fashion. Prior to displacement, women had a central role in the village economy in this region. While men would typically migrate to cities to earn cash, women would remain in the villages managing all household affairs and maintaining their subsistence-level farming. Their activities depended upon access to commons, particularly forests, which were sources of fodder, firewood, and fibre. The relocation site for these villagers gave no access to commons and the agricultural land which they were given access to (but not ownership of) had been mechanised, requiring the assistance of men to operate the farm equipment and generally a different set of techniques, which men were in a better position to be educated about. Losing their connection to agriculture led to the confinement of women to the household, which was experienced as highly disempowering and marginalising. Bisht further suggests that, being more disconnected from public life, women were in a disadvantageous position to adjust to the new routines that their new environment imposed upon them. Resettlement policy contained no programmes for building women's capacity to adapt, and provided no mechanism for the reclamation or compensation for lost commons. Given the fact that women, at least in developing countries, have typically been more dependent upon commons than men for their economic activities (see Agarwal 1994), the situation described by Bisht is likely to be universally experienced. We should also consider the fact that, as Ganguly-Thukral (1996) has noted that women in India are often more dependent upon their connections to community and social networks, where men tend to be more mobile. As such, the fracturing of communities that occurs with displacement is likely to have a greater impact on the lives of women.

The processes of resettlement often also have differential effects on men and women. With reference to Burundi refugees in Tanzania, Daley (1991) shows how the resettlement process can serve to reaffirm and even strengthen pre-existing patriarchal relations within the displaced group. The greater survival pressures being placed on refugees led, in many instances, to an accentuation of the gendered division of labour within the household as a means of consolidating resources. Women were, thus, more enclosed within the household in the context of settlements. This, when combined with the male-centric assumptions of the resettlement and assistance programmes, compounded the problem. Resources were allocated only to men, putting families with female heads of household at an extreme disadvantage. Patriarchy was also amplified by the insecurity of the refugee settlements, as men felt the need to step in and act on women's behalf to protect them from state representatives (Daley 1991: 256). This 'enclosure' of women in turn led to less access to employment and administration, causing a decline in their material condition.

Grounding Displacement: Fieldwork and Ethnography

One of the critical ways to analyse the lived experiences of dispossession is the deployment of the holistic approaches of fieldwork and ethnography. In this regard Oliver-Smith (2009) suggests that the holistic approach of anthropology has made it uniquely suited to documenting the 'totalising' effects of displacement. He notes that anthropologists have been actively involved in almost every stage in the process, including initial consultations, documenting effects, designing more effective policies for governments and transnational institutions alike, working with civil society groups to raise awareness of the effects of displacement, highlighting injustice and identifying weakness in existing policies, and exploring the differential effects of diverse forms of displacement. Anthropologists have been crucial to the formation of fora, including research centres and advocacy groups like Cultural Survival, to draw the attention of the public and policymakers to the issues displaced people are faced with (Colson 2003: 12–14). With development-induced displacement having become a highly politicised issue, with many NGOs and social movements calling for new development models that do not involve dispossession, Oliver-Smith (2009) argues that anthropology has a highly important contemporary role, drawing on its wealth of experience in the field to critique existing frameworks and contribute to the imagining of alternatives.

Ethnographic methodologies have proved invaluable in highlighting the suffering of the displaced and the complex issues they face in resettlement. Colson (2003), in a comprehensive review of relevant studies in the field, shows some of the ways in which these methodologies have proven useful. Longitudinal studies have provided the opportunity for researchers to explore displacement and relocation as a continuous process, with far-reaching impacts hitherto recognised. Ethnography's great concern for particularity and difference has allowed due consideration for spectrum of experiences faced by different categories of people. This recognition of diversity has been complemented with an understanding that being labelled collectively as 'refugees' has concrete effects, and that displacement can give rise to a sense of shared identity.

Ethnographic methods can also give rise to important theoretical interventions surrounding the displaced. Malkki (1995) engages with the epistemological construction of the concept of refugees and the formation and history of 'refugee studies'. She shows that the discipline of refugee studies has often uncritically taken on the assumptions of international institutions such as the United Nations (which often funds this kind of research), which produce the epistemological frameworks through which refugees are constructed for purposes of governance. This can lead not only to an uncritical acceptance of the official narratives on causes of displacement, but also a kind of essentialism which draws a clear, non-permeable distinction between refugees and non-refugees. Anthropology provides a solution to these limitations through detailed ethnographies at the site of displacement. Malkki details, for example, how ethnography can highlight the plight of both those who left their homes and those who did not.[3] Understanding the causes and effects of both positions is critical. Furthermore, ethnography challenges assumptions that, for example, refugees are completely uprooted and disconnected from their homelands, showing ongoing forms of contact and communication between those who left and those who remained behind.

3 This issue has been taken up in greater detail by Lubkemann (2008), who argues that with its focus on the displaced refugee, forced migration studies renders invisible those who are 'involuntarily immobile'.

Multi-sited ethnographic studies provide a different, but highly valuable contribution. Marcus (1995) describes these studies in contrast to more traditional, single-sited ethnographies, in which the focus is on a single location, contextualised within the global system. In multi-sited ethnographies, as it implies, data is collected from multiple locations to examine how ideas and cultural meanings circulate across great distances. It has developed as a response to both theoretical developments in post-modernism and interdisciplinary studies, as well as certain tendencies within the real world, which have led to an acknowledgement that the 'world system' as a contextual frame for local meaning has become problematic:

> [T]he world system is not the theoretically constituted holistic frame that gives context to the contemporary study of peoples or local subjects closely observed by ethnographers, but it becomes, in a piecemeal way, integral to and embedded in discontinuous, multi-sited objects of study. Cultural logics so much sought after in anthropology are always multiply produced, and any ethnographic account of these logics finds that they are at least partly constituted within sites of the so-called system (i.e. modern interlocking institutions of media, markets, states, industries, universities—the worlds of elites, experts, and middle classes). Strategies of quite literally following connections, associations, and putative relationships are thus at the very heart of designing multi-sited ethnographic research. (Marcus 1995: 97)

Through following these connections and associations, multi-sited ethnographies allow an understanding not only of how people's perspectives and experiences are constructed via global processes and systems of meaning, but also the dispersed ways in which those systems of meaning are produced and transmitted, and the role that individuals play within that construction/dispersion. Although Marcus does not talk about the application of this to displacement specifically, the significance of the implicit spatial de-centeredness in his approach, coupled with the recent insights of Hart (2006) with particular respect to the spatial aspects of dislocation we would argue that concerns of multi-sited ethnography is most useful for this field.

Anthropologists have been engaged in this field for several decades now. Colson (2003) notes that this has allowed a sufficiently detailed 'ethnographic base' to develop to allow for generalisations about the experiences of forced migrants. Anthropologists are in a position to make certain claims about likely reactions to displacement, while having enough experience in the field to be open to the ever present possibility of people coming up with new and innovative solutions to their problems.

Additionally, anthropologists have been able to draw on a rich set of theoretical and conceptual tools to understand the processes that displaced people undergo. As Colson (2003) outlines, although in its early history anthropology was mainly focused on more stable societies. In particular, Colson explores how the field of immigrant studies provided several useful tools for understanding the experience of relocation (see also Marx 1990). This literature had emphasised the importance of working through personal networks, using relations of mutual trust and reciprocity in order for migrants to establish themselves in new communities. When applied to refugee studies, this has allowed great insight into the difficulty of their situation. The importance of trust in establishing connections assisting resettlement is a critical issue: trust typically requires a sense of continuing links with others. In the context of refugee settlements, these links are often severed. Normal social life has been interrupted, giving people less social and cultural capital through which to find solutions to their problems.

Introducing the Chapters

This book is divided into three sections. The first set of chapters examines the processes of displacement in South Asia that have resulted from the State adopting neoliberal policy frameworks. All the contributors in this section – Basu, Ganguly-Scrase, Green, Hill and Scrase – highlight the specific consequences of these measures for the poor and the marginalised. A diverse field of forced migrations are considered, including development induced, global tourism-led displacements of local communities of poor youth, gender dimensions of resettlement policies and the intersections of cross-border forced migrations and poverty induced internal displacements.

The following section examines consequences for people stemming from the role of the State in the politics of forced migrations. In these chapters Ahmed, Majid-Cooke and Rekhari pay particular attention to the experiences of minorities and indigenous peoples. The examples range from the experiences of State exclusionary practices against Indigenous children in Australia, particularly their coercive relocation, dispossession of minorities in The Chittagong Hill Tracts of Bangladesh, and the impacts of the State sanctioned land clearing on the Bonggi people of Sabah, Malaysia.

The final section 'Placeless Identities: Non-state places and floating peoples' aims at understanding refugee imaginations across diverse locales. While da Silva, and Julian focus on transnational imaginaries, Lahiri-Dutt and Samanta's accounts are deeply inflected by the local and place-based identities and existence of char dwellers. Ganguly-Scrase and Sheridan bring together intersections of both in that the internally displaced of the Global South struggle to seek asylum in the Global North. The diverse methodologies consisting of interviews, archival research and textual analysis deployed by the various contributors working in different disciplines in this section highlight our central claim that to fully comprehend the nature of dispossession in the new global order requires us to draw on interdisciplinary approaches.

The chapters in this volume can be examined both regionally and thematically. In the following pages we review these overlapping concerns. We argue that while the chapters in the book are divided into sections for the sake of convenience and ease of reading, the complexity of dislocations require us to go beyond outlining each chapter sequentially. For example, Ahmed, Majid Cooke and Rakhari address the role of the state in coercive relocation of indigenous minorities; at the same time, their historical particularities lead to divergent modes of inquiry.

A number of chapters in this volume suggest that the nature of policies pursued by the state leads to various forms of displacement. Ganguly-Scrase and Green specifically examine the consequences of neoliberal policies for different groups in South Asia that has resulted in their dislocation. Therefore a political economy of neoliberal globalisation underpins many of the chapters. However, throughout this volume we make cautionary remarks about simplistic causal links between globalisation and dispossession. Instead we point to the ways in which neoliberal globalisation has accentuated the process of forced relocation. Development-induced displacement and poverty-induced displacement are intrinsically linked. Ganguly-Scrase and Green, and Hill and Scrase specifically address these concerns in their respective chapters on India. Their findings highlight the failure by the state to provide adequate opportunities for survival and ensure social justice to some of its most vulnerable citizens. An important point to emerge from Belinda Green's chapter is the role of the neoliberal state in sponsoring, promoting, and being in service of the private sector. The collusion with power of capital finds its expression not only in economic terms, but also in terms of policing and surveillance.

Hill and Scrase examine the consequences for communities resulting from restructuring of the sea port in the hinterland of Calcutta (now Kolkata). They demonstrate that under neoliberal economic principles, especially those laid out by the World Bank, ports now have to operate more efficiently, attract increased trade and services, provide more streamlined processing of cargo, attract private investment, and make increased profits for private sector partners. Of particular concern is the resultant loss of jobs in local communities and displacement from homes and land as national governments, together with private capital, attempt to restructure and develop these ports in order to meet the demands for a globalised, export economy. Majid Cooke raises similar concerns regarding the indigenous Bonggi people's experience of resettlement in Sabah, Malaysia, which initially took place in the late 1960s. With the establishment of a mega-government-linked agricultural project (rubber plantation), Bonggi are experiencing a new, more rational, and efficient form of development-induced resettlement which could produce future displacement if Bonggi's desire for tenure security is not met with.

Gendered aspects of displacement are also examined in this volume. Basu in her chapter argues that examining the socially differentiated nature of displaced and resettled subjects is central to understanding the gender dimensions of existing displacement processes and the limitations of resettlement sites. Ganguly-Scrase draws attention to gendered complexities of cross-border forced migration and internal displacement by documenting women's shared experiences as refugees and economic migrants. Her chapter demonstrates the boundaries of poverty-induced internal migration and forced international displacements often traverse the same ground.

Market reforms have a negative impact on local communities. The discourse and valorisation of 'choice' is a marker of neoliberal reforms. Paradoxically, however, it is this assumption of choice that is starkly absent in so far as many local populations are concerned, rendering them homeless. Limited support is available due to further withdrawal of the state. While this is amply evident in cases of state–citizen relationship where neoliberal policies have been implemented, in other instances, it is the repressive apparatus of the state is of particular concern. In particular, in the case of indigenous populations, the might of the state is the main problem.

The power of the state over indigenous populations is immense. Rekhari, Ahmed and to some extent Majid Cooke show that both in its colonial and post-colonial forms, the state has played a crucial role in systematically dispossessing them. The process common to the dispossession of indigenous people is the assumption of their 'backwardness'. Therefore, vigorous attempts to transform their pre-modern status can be found in the paternalistic civilizational narratives, which call for state action to pull them up by the bootstraps into modernity. The infantile nature of the inhabitants of pre-literate societies is repeatedly reinforced to deny them the ability to govern themselves. This, coupled with the assimilationist agendas to normalise indigenous kinship practices and cultural forms, have given rise to policies such as forcible removal of children from their families, which Rekhari details in her analysis. Indigenous people's conceptualisation of land ownership is also at odds with their position within modern nation-states. Often it is a common practice whereby a segment of land that already belongs to the original inhabitants is symbolically 'returned' in the form of setting up 'reserves'. This is part of a broader project of enclosed mobility that Ahmed alludes to in his account of the Pankhuas in the Chittagong Hill Tracts.

Border crossing is a reoccurring theme. Ganguly-Scrase's account of women's shared spaces as forced migrants challenges the boundaries of the nation that are inscribed on women's bodies, whereas the importance of alliances across borders are found in the chapter by Ahmed on the Chittagong Hill Tracts in Bangladesh. It is widely known that in the frontier districts, the indigenous 'non-state' peoples in the Chittagong Hill Tracts have had kinship ties across the border with India and Myanmar since medieval times. The politics of identity and displacement of the Chittagong Hill Tracts are caught up in these historical legacies. The conceptualisation of borders provides an important, but not necessarily an exclusive framework for understanding the twin processes of enclosure and mobility. It enables us to critically reflect on the flexible and fluid nature of identities formed in mobility. At the same time, its obverse is shown: the constraints placed on movements that lead to injustice and inequality.

The types of methodologies used have special significance in analysing the experiences of forced migration and resettlement. We see its significance in the works of Green, Ganguly-Scrase, Lahiri-Dutt and Samanta, and Julian, all of whom emphasise the importance of long-term fieldwork and qualitative research methodologies in understanding the plight of the displaced. Lahiri-

Dutt and Samanta highlight the significance of utilizing multi-method approach of qualitative research, which includes observation, in-depth interviews, and document analysis, to interpret the experiences of illegal women migrants in the *char*lands (river islands) of the lower Bengal region, where living environments are uncertain and risky. Here, the illegality of char residents merges imperceptibly with that of the lands on which they live. Their study illustrates women's strategies in coping with the challenges they confront and highlight the problematic issue of non-state forms of geographical contexts of migration.

The power of ethnographic analyses frames the accounts by Green and Ganguly-Scrase in the era of neoliberal globalisation. Green is concerned with the place of youth in displaced communities and the ways in which they have engaged in the process of re-building their lives through the construction of hybrid identities. Belinda Green draws on ethnographic approaches to explore the effects of tourism led displacement in Kerala, India, on a group of Scheduled Caste (SC) male youth who live and work in Kovalam. She shows that in the contemporary setting of the political economy of international tourism in Kerala, is undergoing significant changes as a result of neoliberal policy and praxis. The tourist site of Kovalam in Kerala provides a disturbing exemplar of neoliberalism where privatization, multinational investment, and the development of elite resorts and mass tourism has led to multiple forms of forced displacement and dislocation for local people. Green presents ethnographic account of the ways in which these processes are being met and negotiated by a group of SC male youth. Rather than relying solely on two-dimensional constructions of powerless and marginalized victims, unable to participate in the neoliberal project, this chapter highlights the dynamic and complex response to such conditions by these SC male youth. Through their articulation of a subcultural formation, locally referred to as the *jungee*s, SC male youth seek to engage with and utilize neoliberal forces for their own social and economic mobility.

The notion of agency as an analytical tool pervades in many of the chapters. However, this should not lead us to downplay the powerlessness of the displaced. The concerns raised by Hill and Scrase point to the plight of economic refugees whose conditions are the direct outcomes of development-induced displacement and the failure of the state ensure the basic rights of its citizens. However, given the inequalities that characterize global and national economies and polities, it is unlikely that the outcomes of internal dislocations can be fully addressed within state-based compensatory frameworks that are likely to be geared towards uniformity. Basu's chapter follows this line of argument and focuses on human displacements associated with dam construction along India's Narmada river. It seeks to provide a critical analysis of the ways in which state-led compensation policies have addressed disruptions and reconstructions of social identities and community spaces. More specifically, this chapter highlights the need to understand the difference between land-based and cash-based compensation policies, which have implications not just in terms of changing livelihood practices, but also in terms of new forms of knowledge required to navigate everyday lives in resettlement sites. Such differences become especially pertinent in the context of gender, as women are called on to play new social and economic roles and to rebuild homes in unfamiliar contexts.

Given that migrations and displacements are increasingly being addressed in terms of how such population movements blur national boundaries, internal displacements also illustrate the ways in which fractures within nation-states are heightened through involuntary dislocations, so that the borders that characterize contemporary population movements both transcend as well as reinforce the geographies of social identities. Here, Boktiar Ahmed's account of the experiences of people in the Chittagong Hill Tracts on the margins of the Bangladesh nation-state presents a critical opportunity to explore global-local tensions. The construction of Kaptai dam in 1962 displaced about 100,000 indigenous people. Since then, the increasing Bengali settlement, development projects and industrialization, massive exploitation of land and forest resources, together with violent state control over settlement patterns have made displacement the most detrimental phenomenon of indigenous life. Ahmed's chapter examines the forms of displacement prevailing in the Chittagong Hill Tracts and analyses how a certain political economy of settlement shapes the making of the Chittagong Hill Tracts as margin. He explores the making of a minority and margins in the context of globalization. It draws attention to the disparities in the regions like the Chittagong Hill Tracts which are, at the same time, the margin of globalization itself.

Some authors re-iterate the paradox of globalization in the twenty-first century whereby material and symbolic goods travel relatively freely across national borders, while movements of people, or at least particular categories of people, are becoming increasingly understood as a problem in need of control. Based on research among internally displaced people in Afghanistan and Sri Lanka Ganguly-Scrase and Sheridan examine the perceptions of porous boundaries and unlimited opportunities that coexist in the public imaginary with hardened attitudes towards desperate humans who seek to cross-national borders without authorization. They argue that a preoccupation with national security and border control demonise and criminalise 'undocumented' migrants who arrive without authorisation. Yet, within the contrasting debates between the undesirability of undocumented migrants and advocates of the humanitarian intake of asylum seekers, rarely are the perspectives of those seeking refuge taken into consideration. Therefore, Ganguly-Scrase and Sheridan focus on the intentionality of those seeking refuge.

This volume repeatedly emphasises the urgency of drawing from multiple methodologies and disciplinary perspectives. In this regard, many chapters draw on a mix of documentary films, ethnographic studies, secondary sources, and literary narrative in order to trace the journey from displacement to resettlement through differing modes of approaching personal narratives of the loss and rebuilding of community.

Regional and thematic concerns frequently overlap and intersect. Yet, their analytical insights vary considerably. The chapters on refugees by Simoes da Silva and Julian are an exemplar of this diversity. In both cases, the situation may be read as people's own dealings with the trauma and mass displacement. Their accounts draw attention to the urgency of the need to engage the spirit of an age of post-colonial spaces where the progressive decay of the nation-state, or indeed its near collapse in the form of 'failed states', has come to encapsulate a wider narrative of displacement across the world.

By focusing on the complexity of Hmong identit(ies) in a transnational context, Julian's chapter, first, explores the concept of 'refugeehood' in relation to the Hmong diaspora. Second, it examines how Hmong heritage and identit(ies) are both retained and changed as Hmong people become displaced and dispersed and part of the various local communities in which they have resettled and now live. While Julian's accounts are grounded in fieldwork with Hmong in Australia, the United States, and Thailand, and analyses of Hmong media in the United States, da Silva draws on a range of contemporary literary texts to explore aspects of the representation of the refugee in contemporary culture. Specifically, he highlights the complex ways in which the experience of displacement and often of dispossession lived by peoples perpetually moving across physical and cultural borders supports the views posed by social scientists such as Bauman, Nyers, Agamben, and others. The underlying reason for the focus on literary texts is to explicate their intertwining of a political concern with the meaning of refugees in contemporary society and the way such texts articulate a 'poetics of refugeeness'.

While much has been written about the place of indigenous minorities in post-colonial societies in developing countries, curiously, the forced relocation of indigenous people in white settler societies is rarely examined through the lens of displacement. Suneeti Rekhari fills this analytical gap in her account of the coerced removal of Indigenous children in Australia. She examines the forcible displacement of indigenous people from their lands, homes, and families after the establishment of a penal colony in Australia. In particular, her chapter concentrates on the induced displacement of indigenous children from their families. It outlines some early documented child removal and also the experiences of children taken away under the directions of the government protection and assimilation policies. It refers to existing scholarship in this area and primarily relies on first-hand accounts by children separated from their families, and the accounts tabled at the federal parliament in the May 1997 *Bringing Them Home* report. Rekhari presents a critical examination of the child-removal processes and highlights the lived experiences of forced dislocation. Her ultimate aim is to provide a tool for future analysis for questioning the complexities involved in the 'movement by coercion' of a marginalised section of the Australian population.

We end this brief introductory account of '*Home and Belonging*' with the hope that the critical examination of historical and contemporary accounts of the displaced offered by this book charts a new path in displacement studies. We hope that the grounded case studies offered in this volume encourages a rethinking of the ways in which we have thought about people and their movements from one place to another, across borders and boundaries. These studies show that no matter what might have been the reason of dislocation, human experiences of place and placelessness are similar as subjects recreate their homes and adjust and adapt to new places, and by privileging the human experience, challenge the compartmentalised treatment of displacement that we are becoming accustomed to.

References

Agarwal, Bina. 1994. *A Field of One's Own: Gender and Land Rights in South Asia.* Cambridge: Cambridge University Press.

Ahmed, Sara, Castaneda, Claudia, Fortier, Anne-Marie, and Sheller, Mimi. 2003. Introduction, in *Uprootings/Regroundings: Questions of Home and Migration*, edited by Sara Ahmed, Claudia Castaneda, Anne-Marie Fortier, and Mimi Sheller. Oxford: Berg, pp. 1–25.

Bannerji, Paula. 2010. *Borders, Histories, Existences: Gender and Beyond.* New Delhi: Sage Publications.

Bauman, Zygmunt. 2004. *Wasted Lives: Modernity and Its Outcasts*. Cambridge: Polity.

Baviskar, A. 2005. *In the Belly of the River: Tribal Conflicts over Development in the Narmada Valley* (2 ed.). New Delhi: Oxford University Press.

Binaisa, Naluwemba 2011. African migrants negotiate 'home' and 'belonging': Reframing transnationalism through a diasporic lens, *International Migration Institute Working Papers,* Oxford: Oxford Department of International development.

Bisht, Tulsi Charan. 2009. Development-induced displacement and women: The case of the Tehri dam, India. *The Asia Pacific Journal of Anthropology*, 10(4, December), 301–17.

Castles, Stephen. 2003. The international politics of forced migration. *Development*, 46(3), 11–20.

Cernea, M. 1996. Bridging the research divide: Studying refugees and development oustees, in Tim Allen (ed.) *In Search of Cool Ground: War, Flight and Homecoming in Northeast Africa*, edited by T. Allen. London/Trenton: Africa World Press, pp. 11–55.

Cernea, M. (2003). For a new economics of resettlement: a sociological critique of the compensation principle. *International Social Science Journal*, 55 (175), 37–45.

Chimni, B.S. 1998. 'The geopolitics of refugee studies: A view from the south'. *Journal of Refugee Studies*, 11(4), 350–74.

Chimni, B.S. 2009. 'The birth of a "discipline": From refugee to forced migration studies'. *Journal of Refugee Studies*, 22(1), 11–29.

Cohen, J. and Browning, A. 2007. 'The decline of a craft: Basket making in San Juan Guelavia'. *Human Organization,* 66 (3) 229–40.

Cohen, J.H. 1998. 'Craft production and the challenge of the global market: An artisan cooperative in Oaxaca, Mexico'. *Human Organization*, 57 (1), 74–83.

Colson, E. 2003. 'Forced migration and the anthropological response'. *Journal of Refugee Studies*, 16 (1), 1–18.

Colson, E. 1999. 'Gendering those uprooted by "development"', in D. Indra (ed.), *Engendering Forced Migration: Theory and practice*. Oxford: Berghahn Books, 23–39.

Cresswell, Tim and Peter Merriman 2011. 'Introduction: Geographies of mobilities – Practices, spaces, subjects', in Tim Cresswell and Peter Merriman (eds) *Geographies of Mobilities: Practices, Spaces, Subjects*, Farnham: Ashgate, pp. 1–18.

Cunningham, Hilary and Heyman, Josiah. 2004. 'Introduction: Mobilities and enclosures at borders'. *Identities: Global Studies in Culture and Power*, 11(3), 289–302.

Cunningham, H. 2004. 'Nations rebound? Crossing borders in a gated globe'. *Identities*, 11(3), 329–50.

Daley, P. 1991. 'Gender, displacement and social reproduction: settling Burundi refugees in Western Tanzania'. *Journal of Refugee Studies*, 4 (3), 248–66.

Dwivedi, R. 2002. 'Models and methods in development-induced displacement', *Development and Change*, 33 (4), 709–32.

Escobar, A. 2003. 'Displacement, development, and modernity in the Colombian Pacific'. *International Social Science Journal*, 55 (175), 157–67.

Eversole, R. 2006. 'Crafting development in Bolivia'. *Journal of International Development*. 18, 945–55.

Fernandes, W. (2007). 'Singur and the displacement scenario'. *Economic and Political Weekly*, January 20, 203–05.

Ganguly-Scrase, R. 2013. *Global Issues/Local Contexts: The Rabi Das of West Bengal*. New Delhi: Orient Blackswan, Second edition.

Ganguly-Thukral, E. 1996. 'Development, displacement and rehabilitation: locating gender'. *Economic and Political Weekly*, June 15, 1500–1503.

Gellert, P. K Lynch, B. D. 2003.' Mega-projects as displacements'. *International Social Science Journal*, 55 (175), 15–25.

Ghosh, Kaushik. 2006. Between global flows and local dams: Indigenousness, locality, and the transnational sphere in Jharkhand, India. *Cultural Anthropology*, 21(4), 501–34.

Gupta, A. 1989. 'The political economy of post-Independence India: A review article', *The Journal of Asian Studies* 48 (4), 787–97.

Hart, Gillian. 2006. 'Denaturalizing dispossession: Critical ethnography in the age of resurgent imperialism'. *Antipode*, 38(5), 977–1004.

Hitchcox, L. 1990. *Vietnamese Refugees in Southeast Asian Camps*. London: Macmillan.

Huseby-Darvas, Eva V. 1995. 'Voices of plight, voices of paradox: Narratives of women refugees from the Balkans and the Hungarian host population', *Anthropology of East Europe Review*, 13 (1) 18–33

Hyndman, Jennifer. 2010. 'Introduction: The feminist politics of refugee migration'. *Gender, Place and Culture*, 17(4), 453–59.

Jha, P. 2001. *A Note on India's Post-Independence Economic Development and Some Comments on the Associated Development Discourse,* Institute for World Economics and International Management, URL: http://wiwi.uni-bremen.de accessed on 20 July, 2010.

Kibreab, G. 1999. 'Revisiting the debate on people, place, identity and displacement'. *Journal of Refugee Studies*, 12(4), 384–410.

Kingfisher, C. and Maskovsky, Jeff. 2008. 'Introduction: The limits of neoliberalism'. *Critique of Anthropology*, 28 (2), 115–27.

Knorringa, P. 1999. 'Artisan labour in the Agra footwear industry: Continued informality and changing threats', *Contributions to Indian Sociology*, 33 (1), 303–27.

Korovkin, T. 1998 'Commodity production and ethnic culture: Otavalo, Northern Ecuador', *Economic Development and Cultural Change*, 47 (1), 125–54

Landau, Loren B. and Monson, Tamlyn. 2008. 'Displacement, estrangement and sovereignty: Reconfiguring state power in urban South Africa'. *Government and Opposition*, 43(2), 315–36.

Lubkemann, S.C. 2008. 'Involuntary immobility: On a theoretical invisibility in forced migration studies'. *Journal of Refugee Studies*, 21 (4), 454–75.

Malkki, L. 1995. 'Refugees and exile: from refugee studies to the national order of things', *Annual Review of Anthropology*, 24, 495–523.

Marcus, G. 1995. Ethnography in/of the world system: The emergence of multi-sited ethnography. *Annual Review of Anthropology*, 24: 95–117.

Nolin, Catherine. 2006. *Transnational Ruptures: Gender and Forced Migration.* Aldershot: Ashgate.

Oliver-Smith, A. 2009. 'Introduction: development-forced displacement and resettlement: a global human rights crisis'. In A. Oliver-Smith (ed), *Development and Dispossession: The Crisis of Forced Displacement and Resettlement.* Santa Fe: School for Advanced Research Press, pp. 3–23.

Pallitto, Robert and Heyman, Josiah. 2008. 'Theorizing cross-border mobility: Surveillance'. *Security and Identity in Surveillance and Society*, 5(3), 315–33.

Randeria, Shalini. 2003. 'Cunning states and unaccountable international institutions: Legal plurality, social movements and rights of local communities to common property resources'. *European Journal of Sociology*, 44(1), 27–60.

Said, Edward. 1979. 'Zionism from the point of view of its victims'. *Social Text*, (1), 7–58.

Scalettaris, Giulia. 2007. 'Refugee studies and the international refugee regime: A reflection on a desirable separation'. *Refugee Survey Quarterly*, 26(3), 36–50.

Siu, H.F. 2007. 'Grounding displacement: Uncivil urban spaces in post-reform south China'. *American Ethnologist*, 34(2), 329–50.

Stepputat, Finn. 2008. 'Forced migration, land and sovereignty'. *Government and Opposition*, 43(2), 337–57.

Turton, David. 2004. *The Meaning of Place in a World of Movement: Lessons from Long-term Field Research in Southern Ethiopia.* The Annual Elizabeth Colson Lecture, Rhodes House, University of Oxford, Oxford, 12 May.

Turton, David. 2003. *Refugees and 'Other Forced Migrants': Towards a Unitary Study of Forced Migration*. Paper presented at a Workshop on 'Settlement and Resettlement in Ethiopia', organised by the UN Emergency Unit for Ethiopia and the Ethiopian Society of Sociologists, Social Workers and Anthropologists, Addis Ababa, 28–30 January.

Trimikliniotis, Nicos, Gordon, Steven, and Zondo, Brian. 2008. 'Globalisation and migrant labour in a 'rainbow nation': A fortress South Africa'? *Third World Quarterly*, 29(7), 1323–39.

Tuan, Yi-Fu 1974. *Topophilia: Study of Environmental Perception, Attitudes And Values*, Englewood Cliffs, New Jersey: Prentice Hall Inc.

Walton-Roberts, Margaret and Pratt, Geraldine. 2005. 'Mobile modernities: A South Asian family negotiates immigration, gender and class in Canada'. *Gender, Place and Culture*, 12(2), 173–95.

Wethey, E. 2005) 'Creative commodification of handicrafts, the encounter between the export market and the indigenous weaver: comparisons of Latin American weaving communities', *Lambda Alpha Journal*, (35), 2–28.

Wherry, F.F. 2006. 'The nation state, identity management and Indigenous crafts: Constructing markets and opportunities in Northwest Costa Rica', *Journal of Ethnic and Racial Studies*, 29 (1), 124–52.

Chapter 2

Neoliberal Development, Port Reform and Displacement: The Case of Kolkata and Haldia, West Bengal, India[1]

Douglas Hill and Timothy J. Scrase

This chapter is concerned with examining the ongoing reforms of the global port sector, focusing in particular on Kolkata (earlier called Calcutta) and Haldia ports in India. Over the past decade, many ports within the Asia Pacific have been in the process of restructuring their operations, attracting private capital and rationalising their workforces. This has necessitated changes to labour, port operations, movement of goods, security, and increasing sustainability. Above all, under neoliberal economic principles, especially those laid out by the World Bank, ports now have to operate more efficiently, attract increased trade and services, provide more streamlined processing of cargo, attract private investment, and make increased profits for private sector partners. However, there are numerous complexities and problems inherent in instigating massive port reforms. One of the most significant of these relates to the issue of displacement, since ports often expand on to greenfield sites or to those that have previously been occupied by low-income communities that are then relocated elsewhere. Compared to other infrastructure induced displacements, such as dams (Baviskar 2004; Faure 2008; Judge 1997; Scudder 2005), there has been little critical academic investigation of the consequences of the displacement of people due to the construction or expansion of ports.

This chapter explores some of the issues associated with labour rationalisation and displacement, in particular, by describing and analysing data from interviews and documents concerning the reform of the West Bengal port sector. The social impact of the reforms in Kolkata and Haldia are complex, interrelated, and are felt more deeply by workers who see their livelihoods and conditions fast diminishing. Of particular concern is the resultant loss of jobs and displacement

1 Research undertaken for this project was funded by an Australia Research Council (ARC) Linkage Project Grant (No. LP0348477) awarded to Professor Andrew D. Wells, Professor Timothy J. Scrase, and the industry partner, Meyrick and Associates. Dr Douglas Hill was the Research Fellow appointed for this project. We acknowledge the financial support of the ARC and the Centre for Asia Pacific Social Transformation Studies (CAPSTRANS), University of Wollongong.

from homes and land as the government, together with private capital, attempts to restructure and develop these ports in order to meet the demands for a globalised, export economy. In examining these issues, the chapter argues that the predominant policy towards displacement in the port sector does not adequately consider the ex-post lived experience of those who are displaced. To counter this, it argues for a more significant engagement with sociological and particularly ethnographic approaches to displacement. This is in consonance with a range of authors concentrating on forced migration that have argued for an approach that recognises similarities between refugees and persons internally displaced by poverty, noting that the economic and political distinctions made to these different categories of persons often understates the extent to which their post-displacement lived experiences are similar (Binder and Tosic 2005; Chimni 2009; Colson 2003; Ganguly-Scrase and Vogl 2008).

Displacement and Large-scale Infrastructure

Many projects throughout the world that involve large-scale infrastructure, such as dams, airports, and power projects, are premised on the displacement of large numbers of people. Dwivedi (1999) has argued that the enormous volumes of literature that analyses these displacements can be characterised according to a two-fold typology in terms of their approaches. On the one hand, a managerial approach has become common, which assumes that displacement is merely an inevitable but regrettable part of development. All too frequently, this approach is confined only to an identification of possible costs and benefits, so that the debate between advocates and critics becomes confined to an analysis of acceptable numbers, in the process seeking only to effectively manage affected populations. To critics of this approach, the discursive construction of project-induced displacement becomes centred on the appropriate mechanisms to quantify who are the legitimately aggrieved populations and then to dispense best practice compensation that effectively rehabilitates these populations.

On the other hand the 'radical-movementist' approach, which is often subscribed to by civil society organisations, argues that displacement is indicative of a crisis in the way that development is approached. Almost 20 years ago, Claude Alvares (1992) used the medical analogy of *triage* to describe the process by which certain groups of people were deemed to be a dispensable consequence of the development process, a prognosis that led him to argue that development as a project of modernity was fundamentally flawed. An example that those such as Alvares might point to as evidence of the inherent 'violence of development' is the recent Manila North Harbour project, which civil society activists claim would displace more than 800,000 people despite the fact that the project documents assessed those numbers, who were legitimately eligible for compensation for their displacement to number only in hundreds. The Philippines is far from an isolated example of the structural violence of such development, since, as we will examine later, the adaption of contemporary models of port organisation and operation in most places in the world required under neoliberal globalisation usually involves the displacement of significant numbers of people.

In spite of widespread opposition by civil society organisations to displacement due to port relocation or reorganisation, studies of port-induced displacement in academic literature are rare. In the Indian context, the recently inaugurated Gangavaram port in Visakhapatnam, which will displace 25,000 fishermen in those areas and the adjoining village of Dibbapalem, as well as the POSCO integrated steel, mining, and port project in Orissa, has led to considerable agitation amongst civil society groups. In spite of this, there has been little academic interest in the port-related social impacts, with the exception of Parasuraman's examination of displacement induced by the construction of JNPT port in Mumbai. It discusses port-induced displacement when 2,584 ha land from 12 villages in Raigad district adjacent to Greater Bombay district was acquired by the government of Maharashtra (Dwivedi 2002; Parasuraman 1996, 1999). His study, carried out five years after the port was commissioned and three years after rehabilitation work was completed, found that 91 per cent families that had previously owned land had lost all of it, but less than 33 per cent of those families had at least one member of the household who got employment. Parasuraman uses the case study of JNPT as part of a broader argument that suggests that development in India is predominantly characterised by lopsided benefits so that the majority of gains accrue to an elite and relatively small proportion of the population. To him, the consequences of port-sector restructuring are indicative of broader processes that explicitly endorse a structural bias, echoing much of the writing in the literature on Indian political economy that argues that there are inherent class biases in the development process (Bardhan 1984; Corbridge and Harriss 2000). Ganguly-Scrase and Vogl (2008), among others, have argued that the manner in which both refugees and the internally displaced are discursively constructed elides significant gendered aspects of displacement and reflects a patriarchal context. Similarly, work on other instances of forced migration has argued for a need to examine both the structural conditions that facilitate and legitimise displacement as well as examining the ex-post social realities of those who are displaced.

This chapter accommodates these approaches in the following manner. Before proceeding with analysing aspects of port development induced displacement, some background to the Indian port sector and the globalization of ports and shipping in the era of neoliberal development is presented. Importantly, defining 'what is a port' is not a simple question as a port has far reaching social and environmental influences on both land and water. Defining a port, and understanding its surrounding areas of influence and impact, its hinterland, thus, raises significant and complex economic, political, and social issues at local, regional, and national levels. The middle section of the chapter provides an outline of the specifics of the development of Kolkata and Haldia ports, while the latter part of the chapter draws on qualitative interviews conducted with port managers, unionists, and especially draws on interviews of those who have been displaced by the construction of the facilities at Haldia. Importantly, the port sector exemplifies significant trends in the way that populations under neoliberal economic transformation are managed and, thus, greater attention to the voices of those displaced needs to be found.

India's Ports: A Brief Background

Ports are an important element in India's trade, since over 90 per cent of the volume and 70 per cent of the value of India's trade is through maritime transport. In the early twentieth century, India had only five ports (including Karachi) primarily responsible for the majority of cargo coming into the Indian subcontinent. In the post-Independence period, this eventually increased to 12 major ports which, in the current era, handle 75 per cent of all traffic (Figure 2.1).

Each port is administered by the Central Government through the Indian Ports Act 1908 and the Major Ports Trusts Act 1963 and usually headed by an Indian Administrative Services (IAS) officer. A major source of revenue is tariffs, the ceilings of which are the responsibility of Tariff Authority Major Ports (TAMP). Labour in these major ports is regulated through the Dockworkers Act 1970, which has created a pool of labour from which the workforce must be drawn. Dockworkers are employed in gangs, the size of which varies significantly between ports.

In addition to the major ports, there are around 200 minor or intermediate ports, run by state governments or by the private sector (Department of Economic Affairs 2009: 2). Of these, only around one-third handles cargo, often specialising in niche cargo, such as dry bulk, liquid bulk, and break bulk. When tracing the relative share of different kinds of ports in India, it is important to note that the demarcation between major and minor ports is less about the size of cargo handled and more about which level of government has jurisdiction over their operation.

One aspect of the differences between major and minor ports relates to management and organisational structure. The major ports remain trusts and are subject to control by a heavily bureaucratic, managerial culture that developed in the colonial era. In contrast, minor ports arguably have more latitude in their operations, since they are able to employ outside agencies or individuals who have experience in port management from elsewhere. Furthermore, minor ports are not subject to price regulation by the TAMP and do not have manpower issues that major ports, many of which still utilise collective labour drawn from Dock Labour Boards. There is, therefore, arguably greater latitude for greater private sector involvement in minor ports.[2] In contrast, most privatisation in major ports

2 An example of this is Pipavav, in Gujarat, which has been running for many years as a fully privatised port operated by APM Terminals. Pipavav Port was set up by Gujarat Pipavav Port Ltd (GPPL). Several private parties, such as Maersk Shipping, acquired a stake in GPPL, which is a joint venture between SeaKing Infrastructure and the Gujarat Maritime Board.

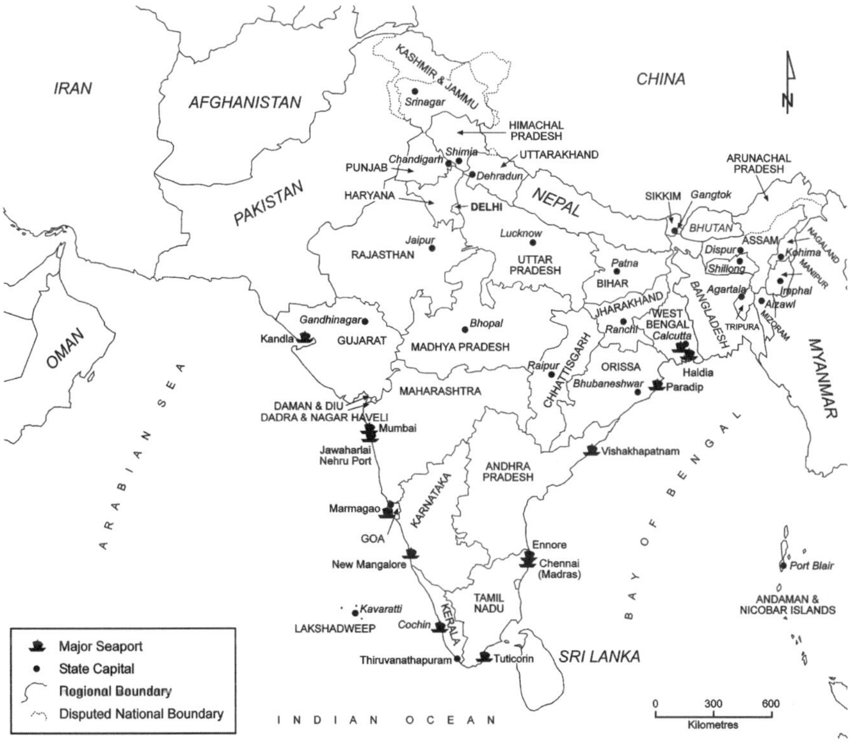

Figure 2.1 Major Seaports in India

has been only a shift towards a partial landlord model with private sector investors leasing berths. In the case of minor ports, additional finance would come from state governments, many of which have been facing their own financial difficulties.

Port development in India is, thus, seen as crucial in building the export-led economy, as well as facilitating much desired and needed imports. In terms of ports, in the Tenth Five Year Plan (2002–7), the Government of India earmarked the following areas as suitable for private sector participation (Planning Commission 2002: 968).

1. Leasing out assets of the ports.
2. Construction and operation of container terminals, multiple cargo berths and specialised cargo berths, warehousing, storage facilities, tank farms, container freight stations, setting up of captive power plants, etc.
3. Leasing of equipment for cargo handling and leasing of floating rafts from the private sector.
4. Pilotage.
5. Captive facilities for port based industries.

In order to facilitate the increased involvement of the private sector the Government of India's Committee on Infrastructure has suggested the relaxation of regulation so as to allow 100 per cent Foreign Direct Investment (FDI) under the automatic route for port-development projects. Accompanying the expansion of the port sector has been the rise of Special Economic Zones (SEZs) many of which will be situated adjacent to ports and will draw investment from private sector companies, state owned and run Public Sector Units (PSUs), state governments, and foreign investors.

Asian Ports in an Era of Globalisation

The dramatic growth and transformation of Indian ports must be seen in the context of the global revolution in shipping and ports over the past three decades. The most significant development has been changes in technology associated with the transportation and distribution of cargo, particularly the widespread introduction of containerisation since the 1970s. This process had a dramatic effect on the composition of maritime trade from the 1990s onwards, as world port container traffic doubled between 1990 and 1998 and has continued to grow exponentially ever since. Between 1990–99, world container trade grew at an average 9.1 per cent, between 2000–09, 7.3 per cent, and is forecast to increase by another 5 per cent by 2015 and whereby it is estimated that 70 per cent of world trade is currently containerized (UNESCAP 2005: 28–9). There is heavy concentration in shipping ownership and container trade monopolisation as indicated in Table 2.1, where the top 10 lines control almost 60 per cent of market share of global container trade (or TEU).

Table 2.1: Top 10 Container Shipping Lines (2009)

Rank	Operator	TEU Capacity	Market Share (%)	Ships	Market Share (%)
1	APM-Maersk	2,031,886	15.5	539	8.91
2	Mediterranean Shipping Company	1,469,865	11.2	425	7.03
3	CMA CGM Group	988,141	7.5	378	6.25
4	Evergreen Line	624,536	4.8	176	2.91
5	Hapag-Lloyd	488,135	3.7	128	2.12

Rank	Operator	TEU Capacity	Market Share (%)	Ships	Market Share (%)
6	COSCO Container Limited	485,796	3.7	148	2.45
7	APL	473,170	3.6	131	2.17
8	CSCL	450,928	3.4	143	2.36
9	NYK	433,000	3.3	119	1.97
10	Hanjin / Senator	378,282	2.9	91	1.5
World Fleet TEU Capacity		13,108,859	100	6,048	100

Source: http://gcaptain.com/the-ten-largest-container-shipping-companies-visualized?678 [accessed: 7 October 2011].

The rapid growth in container traffic as the premier form of maritime trade is primarily due to the reduction in freight rates, which is in turn connected to the economies of scale that have occurred with the growth in the size of container vessels. However, since these economies of scale substantially disappear if the vessel's capacity is under-utilised, the process has also been associated with relentless competition among major shipping lines (Turnbull 2001). The most dramatic effects of containerisation have been reduced ship turnaround time, which has massively reduced labour costs (ILO 1996). Not only has this transformed the efficiency of distribution and enabled goods to be handled more quickly, it has also meant an increase in intermodal transport as containers can be transferred from ship to rail, air or road with greater ease and efficiency. These changes in the shipping and ports sector have had a significant impact upon the importance of export competitiveness between nations, which has itself increased with the global liberalisation of trade. Geographically dispersed countries are now able to compete with each other on a cost basis since transportation has become less of a barrier to entry in the market than at any time previously.

Occurring concurrently with the reduction in freight cost has been an increased concentration in the ownership of shipping lines and stevedoring operations. As Table 2.1 demonstrated, the majority of global maritime trade is now controlled by an oligopoly of shipping lines and container operators who have consolidated their position through a series of mergers. For these operators, this has meant more power in negotiating concessions and facilities from individual ports. Consequently, ports are constantly competing with regional rivals to attract these container liners on the basis of higher productivity and lower costs.

The increased efficiency and reliability of containerisation, in combination with improved logistics and information technology, have had an important effect on manufacturing. These developments have enabled a shift to 'just-in-time' methods, which have increased the flexibility and responsiveness of industry.

Increasingly, then, countries seeking to develop export industries have been forced to transform their operations so that they conform to global standards of business and trading practices. The changes in transport, liberalisation of trade, and manufacturing have compelled nations to increase productivity and reduce costs in their respective port sectors (ILO 1996: 3).

Just as maritime trade has become increasingly concentrated in fewer corporations, the proportion of trade has also become restricted to fewer cargo ports.[3] This increased concentration reflects that some ports have become global feeder hubs, while others remain regional hubs. The intensity and efficiency of these links varies between regions, and there is an increasing regional competition between developing countries to capture transhipment traffic by evolving into a hub port. As such, there has been increasing pressure on ports of all sizes to modify their operation to be compatible with the needs of global capital. This process has necessitated port reform, guided largely by the prescription of multilateral development institutions. The evolving consensus on the most appropriate management structure today is the *landlord port model*. In this model, 'the port maintains ownership while the infrastructure is leased to private operating companies' (Brooks 2004: 170).

In this situation, port authorities have had to both invest in infrastructure to accommodate the new dimension of maritime trade (through measures, such as channel digging, upgrading cranes, logistics, and stacking systems) and reduce costs as much as possible (Turnbull 2001). The large capital outlays needed for the transformation from a multipurpose berth to a container terminal have proved less problematic for industrial countries and Western conglomerations than they have for many developing countries (ILO 1996:4). Consequently, much of the infrastructure in the world's ports is now owned and operated by a small oligopoly of interests, many of which also own and control shipping companies. In these circumstances, ports are under pressure to compete globally with other ports, which has implications for labour relations and infrastructure.

Developing the Port of Kolkata

The early history of the port of Kolkata (formerly Calcutta) is intimately linked with the story of British expansion into India. Prior to the entry of the British East India Company, much of the maritime trade was centred on the western coast, in ports such as Surat (in present day Gujarat) or in the south-eastern region of the Coromandel Coast. With the expansion of the East India Company into Bengal to acquire cotton muslins, raw silk, and saltpetre, Kolkata Port usurped the ports of Satgaon, Chittagong, Hugli, Anjaner, Jaleser, and Balasore as the north-east coast's

3 According to the International Transport Workers' Federation (ITF), although there are more than 2,800 international cargo ports, 80 per cent of total sea-borne trade is handled by just 40 of these ports (ITF 2004: 1).

main port (Kidwai 1989: 209–10). As British presence in the region grew in the late nineteenth century, the port expanded with the addition of many berths. By the 1920s, the economy of both Kolkata and its hinterland were growing; hence, the volume of trade at Kolkata Port increased. Like most of the smaller ports that it replaced, Kolkata is a riverine port, located 221 kilometres from Sandheads; to reach there, one has to travel 148 kilometres of the total distance by river.

In the contemporary period, the development of a new port complex at Haldia (1980–81) has dramatically changed the port scene in West Bengal. The newer facilities at Haldia were designed to facilitate the expansion of Haldia Petrochemicals, which has subsequently developed into a significant industrial cluster. The success of Haldia has meant that most of the growth of traffic into West Bengal has come through this port rather than through Kolkata. This has led to a management scenario where the two ports are run separately in all matters although they remained formally subsumed into the Kolkata Port Trust (KoPT). In the future, all projections point to the continued growth of traffic at Haldia, while Kolkata will continue to stagnate, exacerbating tensions over the bifurcation of the Port Trust into two separate entities.

One aspect affecting the form of Kolkata's transformation is the issue of draft, since the 6–6.5-metre draft at Kolkata prevents the entry of larger, newer ships. This means that the transfer of cargo to Kolkata inevitably involves smaller ships or the unloading of ships from Haldia so as to raise the draft of the ship by lightening the load. Although the draft clearance at Haldia is better than that at Kolkata, it is far from sufficient given requirements of international shipping. There are now moves to put greater emphasis on a site a Saugor Island, which is near the entrance to the Bay of Bengal and boasts a 50-metre draft. Figure 2.2 shows the location of the various ports important to understanding the contemporary situation in Kolkata. The old port is located at Kolkata itself. The newer facilities of this port were developed at Haldia. Recent moves have seen a shift of some 'mid-stream' activities to Diamond Harbour. Further developments at Saugor Island will substantially reduce the draft problems that the port suffers from. Potential competition to the port may come in the future from the development of a private port at Kulpi.

Another crucial difference between the ports of Haldia and Kolkata is the labour regimes under which they operate. Kolkata retains a DLB from which regularised workers must be drawn, while Haldia's workforce is a non-unionised labour force drawn from the external market. In West Bengal, industrial stagnation, particularly from the 1960s, led to a decline in traffic at Kolkata Port, which, in turn, prompted calls for amendments to workplace practices and legislation in 1970. This set the benchmark for conditions which have prevailed amongst registered DLB members ever since.

Ruled for 34 years (1977–2011) by a Left coalition government (the Left Front)[4] until 2011 when a new government formed office, the West Bengal

4 The leading party of the Left Front was the Communist Party of India (Marxist) or CPM.

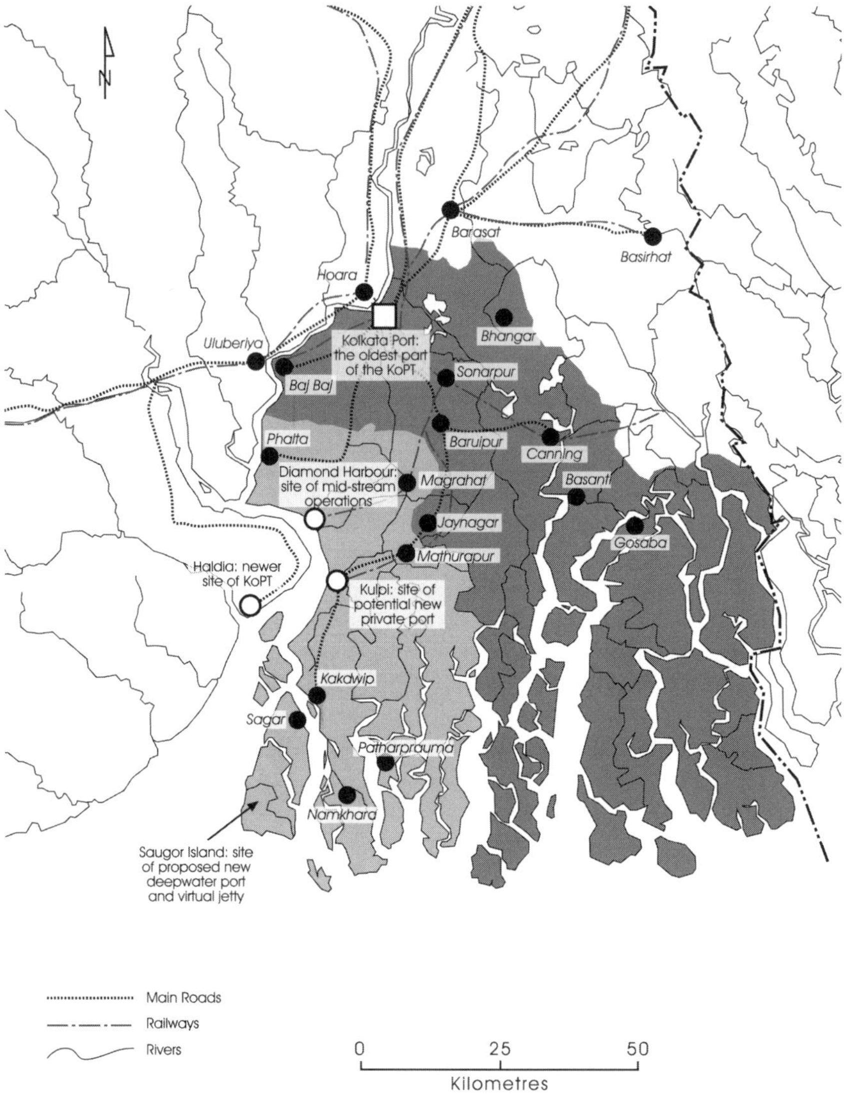

Figure 2.2 Present and Future Ports in West Bengal

economy nevertheless continues to face uncertainty, particularly after India began to liberalise its economy in 1991. The clearest success for the government has been the growth of an industrial cluster around the Haldia Petrochemicals Plant, which adjoins the site of the Haldia Dock complex. Since this time, Haldia Dock has increased its proportionate share of the total volume of trade coming into the Kolkata Port and has been at the forefront of West Bengal's industrial revival.

Although the Port Authority is still responsible for the Port's operation, there have been moves in recent years to shift to a partial landlord model with attempts to increase private-sector investment. An ongoing struggle over delinking and corporatisation frames the relationship between Kolkata and Haldia. Port authorities at Haldia would like to see the delinking of Haldia from Kolkata, formally as two separate entities. In contrast, authorities at Kolkata Port view the continuation of the relationship as a part of an integrated strategy for trade facilitation throughout the state and believe that the separation of the two ports would irreparably weaken the financial position of Kolkata.

As the KoPT attempts to increase its competitive position, a pressing problem is the need for labour restructuring. The DLB system still exists at Kolkata, even though most other major ports in the world have phased-out this system of pooled or collectivised labour. Reasonably good wages, good working conditions, and retirement benefits prevail for DLB workers. However, the conditions available to the next generation of workers are not as lucrative or as secure. The dramatic reduction in labour force from 1969, when there were 43,000 port and dock workers, to the contemporary period, when KDC and HDC together have around 10,000 workers and the DLB less than 1,000 registered workers, has been accomplished incrementally. A factor contributing to the reduction in dock labour force is that the workforce is now ageing and non-DLB workers are more frequently engaged in the work. The port authorities believe that a workforce of 2,000 to 2,500 is sufficient for the KoPT, including both Haldia and Kolkata. Financial problems for the KoPT are ongoing and escalating. The KoPT continues to support approximately 32,000 pensioners. From interviews with port managers, the financial problems that this is causing the port authorities are occurring because approximately 1,000 pension eligible workers reach retirement age every year. Similarly, managers also revealed that, while the DLB had previously received income from a levy system, the withdrawal of this system has caused an acute funding crisis. For example, already burdened to pay pensions to 7,000 registered workers, the Board failed to pay wages to workers from August 2004 until February 2005.

One solution to the potential financial ruin of the KoPT is to outsource their assets, principally land and buildings in Kolkata. The KoPT would like to encourage the re-development of this port land for industrial purposes, including SEZs for industries, such as ready-made garment processing, which already has a large presence around the port area. The use of short-term leases is preferable as it is estimated the leaseholders (including state government departments) of KoPT properties (warehouses, land, and other buildings) owe the KoPT approximately Indian Rs 840 crore (US$170 million).

Another point to note is that Kolkata and Haldia ports are important in the strategy for the development of eastern India, which is reflected in the fact that National Maritime Development Programme (NMDP) foresees an additional combined commitment of Rs 4,723 crore to both ports over the next 10 years. The major emphasis is on construction to augment capacity, both in the development of new facilities and the upgrading of pre-existing facilities. Again, Public-Private

Partnerships are emphasised, with the public sector being responsible for common use infrastructure and the private sector to finance and run those areas that are predominantly for commercial use.

Displacement/Resettlement and Port Development

> The lack of proper dredging and not global warming is behind the obliteration of two islands in the Sunderbans, if experts are to be believed. *Around 10,000 people were rendered homeless when the Lohachara island vanished from the map in the late 1990s.* There is, however, some good news: this island, the first inhabited one in the world to get obliterated is emerging again (Ray 2009; emphasis added)

In the introduction to this chapter, it was indicated that the social impact of ports are far reaching and significant. The immediate local and as well as surrounding areas of the port hinterland (which may in fact be many kilometres distance from the actual port site) are affected to varying degrees by the development or re-development of a port. This is the case with Kolkata Port, particularly its Haldia operations. Research with Kolkata Port employees – management, unionists, and labourers – was conducted in 2004/05 by Douglas Hill. Although the research did not focus solely on the issue of displacement as its main focus was the social impact of port development; more generally, there were nevertheless significant instances and types of displacement mentioned in the interviews. The foregoing discussion mentioned the financial difficulties of the KoPT, the government's agenda to rapidly develop, upgrade, and rationalize port facilities, and the full or semi-privatization of India's ports, all of which is happening under an increasingly neoliberal development agenda. While there have been financial gain and business opportunities for the lucky few, within the Haldia area, there has been widespread displacement, dislocation, job losses, and a range of other social and environmental impacts felt largely by the marginal, local population of peasant landowners and contract, migrant (from other districts) labourers.

Illustrative of the impact of port development on local areas is the case of Nandigram, which is a rural area in East Midnapur district of West Bengal, India. It had been the centre of peasant resistance against an attempt by the West Bengal state government to acquire agricultural land for setting up an SEZ (and the proposed TATA Nano motor vehicle plant which was eventually scrapped in 2009). Nandigram is located around 150 kilometres from Kolkata, on the south bank of the Haldi river, a tributary of the Hooghly river, opposite the industrial city of Haldia. The area falls under the Haldia Development Authority.

The continued requirement for dredging of the channels for large shipping means the eventual financial and time pressures to dump the silt on the immediate shoreline adjacent to Nandigram. In any case, indirect silting is already a significant environmental problem along the shoreline due to the daily dredging of the channels.

This impacts the local fisheries and access to the sea due to shipping traffic. Additionally, pollution, shipping waste, and several other environmental concerns associated with large scale port and shipping activities are major ongoing problems.

Labour Relations

Undoubtedly, one of the most contentious and potentially problematic areas of port reform is labour restructuring. Indeed, since dock workers wield such disproportionate power in affecting global trade, the long history of strike waves in many countries in response to technological change has meant that ports have long carried a reputation as a particularly difficult area for labour restructuring. The 1990s, in particular, saw a rapid increase in strike action by dock workers in many parts of the developed and developing worlds as a response to the rapid changes in the sector (Turnbull 2001).

While organised labour has traditionally been strong in Kolkata Port, in recent years there has not been significant action against the changes in labour conditions and dock labour workforce. Indeed, key union leaders, affiliated to the CPM and Communist Party of India (CPI), backed port trade unions, such as the Calcutta Port & Shore Mazdoor Union, the Calcutta Port and Dock Industrial Workmen's Union, and the Calcutta Dock Workers Union, suggested in the interviews conducted for this research that the new competitive environment, built on partial privatisation, makes these changes inevitable. Instead, they hope that the new infusion of capital and augmentation of existing capacity, that increased private sector involvement should entail, will translate into increased security and prosperity for their members. In reality, however, the major initiatives outlined under the NMDP suggest that their influence will continue to diminish as more and more activities are relocated to purpose built facilities geographically located outside the jurisdiction of the DLB.

There are a variety of scenarios likely in labour restructuring of the port sector. For those who remain employed, the technological change implied in containerisation may have benefits for labour, since increased predictability that comes with containerisation should lead to decreasing casualisation as the labour requirement can be precisely planned. Furthermore, the increases in productivity expected from port restructuring should translate into increased wages and a re-invigorated sector. This should, in the medium term, increase employment opportunities, though usually nowhere near pre-reform levels. However, workers indicated the difficulties of their jobs in Haldia. The four vignettes that are presented now indicate the ongoing financial and daily struggles for these dock workers. Three of the four indicate the necessity for them to have joined the CPM to secure this dock employment. Hiralal Das Adhikari (aged 40) says:

> I am a commerce graduate. I stay at ward of 13 of the Haldia municipality. I have been working as a contract labourer in Haldia dock for the past seven years.

> Earlier I worked in a non-banking chit fund and in the Indian Oil Corporation as a contract worker. I have been looking for a job of primary school teacher for years, but received no call from the employment exchange. I have become a member of the CPM party. It is my contact with the party, which helps me to earn the livelihood now. My two children are studying. However, it has become increasingly difficult to support their education with my uncertain and meagre income. I get only six to seven days employment in a month now. There is lot of money in the hands of only some sections of people. Here, price rises are substantial. However, people like us are at their wit's end on how to cope up with the spiralling price rises and struggle to feed our family.

Sanatan Parua (aged 45) comments:

> I have passed *Madhymik* (school leaving exam). I stay at ward of 16 of the Haldia municipality, a rural periphery of the town. I have been in touch with the CPM party for many years and earned party membership in 1999. I used to earn Rs 1,500 per month as loader–unloader at a private warehouse. After joining Five Star [a labour company], the dock job has increased my income. However, it is not at all sufficient to support my family. My children have stopped going to school since I cannot afford it anymore.

Sk. Mansur Ali (aged 36) has this to say:

> I used to earn around Rs 100 per day by running van rickshaw, a local semi-mechanized goods cum passenger carrier. Membership of the CPM party helped to get the registration in the Five Star labour pool. However, ongoing financial crisis impairs my children's education progress.

And Asoke Das (aged 40) mentions:

> I have passed fifth standard, am married, and have three children. I used to do sun-plastering and painting. Also served as contract labour in the Indian Oil Corporation and earned Rs 3,000–4,000 per month. However, that job is no available anymore. Association with CPM led to registration in Five Star. We do not get more than 12 days jobs in a month now since break, bulk cargo has dwindled and container cargo has increased. I am anxious about the future of my children.

So, along with massive layoffs, the conditions demanded by the corporations which control the world's ports and container lines often preclude the unionised culture of traditional dock labour. There are many cases around the world where this is achieved by directly employing a permanent labour force and by demanding that much of the new labour force is drawn from outside the traditional dock labour (World Bank 1995: 29). Thus, on the one hand, gains in productivity may lead to better conditions for workers in some ports, particularly in the developed world, while on the other hand, elsewhere it is possible that the reduction in collective bargaining power may leave workers more vulnerable without a concomitant increase in working conditions. Given the inevitable costs of restructuring, a preferred strategy is to phase-out the workforce by voluntary retirement. When this is not possible, or is insufficient, ports often opt to employ workers casually or give them reduced hours to ease the transition process. Social funds and poverty alleviation programmes are now commonplace as a mechanism to ease the dislocations and structural unemployment resulting from adjustment and World Bank port restructuring finance sometimes includes these kinds of measures (Graham 1994).

Land Acquisition, Dispossession, and Resettlement: Narratives of Displacement

> The biggest port user industry is Haldia petrochemical, which has 800 direct employees, transport and other 2000 additional. *The local economy thrives on the port and the standard of living has gone up.* Villages around Haldia have their bread-earners in the port town. We have given jobs to one person per family who had lost their homes and land due to the dock complex. Displaced families were rehabilitated further inland. Haldia port's job generation ability is now better that Kolkata. (M.L. Meena, emphasis added)

This quote, by Mr M.L. Meena (the then Deputy Chairman, Haldia Dock Complex, KoPT) epitomises the generally positive views of Haldia from those who manage the complex. They feel the process of rehabilitation and job creation has gone well over the past two decades, despite the realities as expressed by several workers interviewed. Even an administration manager at Haldia, in a 2005 interview, expressed the following:

> The port is sprawled over 67 mouzas or villages. Altogether, 1440 families have been displaced due to land acquisition. Among them about 600 families are yet to be rehabilitated. We will hand over 52 acres of land for the rehabilitation. A Rehabilitation Advisory Committee looks after the process. Since we could not provide alternative land to all the displaced families, many of them are squatting on port land, which originally belonged to them. Rehabilitation means the port authority would provide them homestead land plus civic amenities such as roads, electricity, school etc. The CPT or KoPT is committed to develop the land and amenities. 200 acres by the state government was given free. 52 acres, originally meant for rehabilitation, have been used for industrial development by the state government. Government buildings, residential areas for government staff and Haldia Development Authority, and various shops and markets were built on the land earmarked for the rehabilitation. The state Government wanted

> to create the social infrastructure for industrial development. They had their own compulsions. We will give the state another 52 acres, but this time not free of cost. *The original evictees have legitimate grievances. Once we will be able to handover the extra land, we will vacate the encroachment on our land. Encroachment on port land is a big problem.* Better coordination among the state and central government as well as the port is needed. Local self-government body is in the hands of the pragmatic leaders. They have control over both the dock workers and local people. Haldia is a zero-strike port. *Labour relations are very good.* (Emphasis added)

Indeed, relocation was not a simple process of one getting a fair payment for compulsorily acquired farm land. A certain amount of timing and luck was involved, particularly with the immense increase in land values over the past decade. Himanshu Das, secretary, Haldia unit, Calcutta Port & Shore Mazdoor Union affiliated to the Centre of Indian Trade Unions (CITU) remarked:

> Land acquisition began in 1965–66 over 67 villages. More than 1000 families were displaced. Among them 300–400 families are yet to be rehabilitated. One person from each of the evicted families has been given a job in the port. Many of the families had left Haldia and crossed the river to South 24 Parganas [a district in greater Kolkata] to settle there. Some of them did not realize that the land premium would shoot up here with ongoing urbanization. Therefore, they sold off the piece of homestead land they had received from the port to the outsiders. Now they regret their decision. Some others did not pursue the complex settlement and compensation process. Now they are in the soup [sic]. The amount of compensation was meagre. It was Rs 1070 per bigha. Those had lost the home; they were supposed to get 4–6 decimal homestead lands per family.[5] However, many of the evicted families are yet to receive the alternative agricultural and homestead plot till today. The increasing scarcity of land has compounded the problem. The displaced families are earning their livelihood by running small business, doing contract jobs even as handcart drivers.

In interaction with the displaced persons and people in various settlement colonies of Haldia, different stories emerge. Two case studies illustrate the issues. Nirapada Mondol, a former CPT office superintendent, relates the following:

> The name of our original village was Giraberia-Samraichak. There were 150–200 people in our village when the land was acquired in 1962–63 for the upcoming port. Our family had 30 bighas of land including a homestead and agricultural land. The government acquired all the land. We received Rs 2,645 per acre as compensation [and] also got 12 decimal lands at Durgachak

5 bigha = 1/3 acre; decimal = 1/100 acre; crore = 10 million; lakh = 100,000; 1 US$ = @45 Indian rupees

> rehabilitation colony. The government charged Rs 1,000 for processing the settlement land. The colony was developed for 50 people first. Now it has about 800 plots. Initially six decimal plots were available at Rs 250 only. Thirty years back, nobody thought [the] present level of mass urbanization would happen. It was almost a wilderness here. Many displaced families sold off their piece of land and left for other places. Thirty to forty per cent of resettlement plots have changed hands during the last 25 years. Many succumbed to the pressures from the moneyed outsiders and influential locals and sold off the land at premium value to invest the money in procuring agricultural land elsewhere. The land value here at Haldia Durgachak area has skyrocketed, as it is Rs 20,000–30,000 per decimal now. *Those who had earlier sold their piece now regret their decision.* Legally, the sale was not tenable. However, the municipal authority is pragmatic as they sanctioned building plans, gave water and electricity connection, but are yet to recognize the change of ownership. It leads to cobweb of complex litigations and complications. Initially, we did not regret the land alienation and transfers as much as displaced family's jobs. However, our children are jobless now. (Emphasis added).

Bhupaticharan Dutta is primary school teacher. As he explains:

> My village was called Taramanicahk. We had three and half acres of land, which were acquired in 1965. We received Rs 2,300 for swampland and Rs 2,650 for homestead land. Now the colony land fetches Rs 50,000 per decimal. None of my family has jobs since I had a teaching job. However, many others bagged the port jobs despite having a regular source of income, courtesy of their contacts and clouts in the ruling party [the CPM]. I built a home on my piece of land in the resettlement colony. There are pressures to sell off that land and building which I have been resisting. However, many people could not. In my bloc, there are 10 plots. Four of them have been sold off to outsiders, non-displaced people. These new settlers are port workers who had come here from outside or businessmen. *Those who have good connection in the ruling party or powerful politicians can easily manage to take possession of the resettlement land, even if illegally.* It seems we are destined to be uprooted for a second time in our life. (Emphasis added)

And finally, one government staff at the Durgachak office in Haldia who works on resettlement issues says:

> Many of the original displaced families have not received full compensation since they had left Haldia in search of a new livelihood. *Those who had sold off their homestead rehabilitation land at a throwaway price now want to get it back.* This has happened not only in Durgachak, but at other rehabilitation colonies in Hatiberia, Bhabanipur, Ramnagar, and Debhog. *Prime plots have been transferred or sold to outsiders at the rate of Rs 30,000 to 1 lakh [Rs*

> *100,000] depending upon the locations. The transfer has been illegal earlier but now the administration is regularizing it, as there is pressure from influential quarters*. A substantial portion of Durgachak colony plots have been transferred. It was Rs 250–300 per decimal land in the 1970s. Now it will fetch Rs 1 lakh. In 2000, one acre of land officially sold at Rs 7 lakh in Basudevpur area. *Displaced persons would be pushed back further inland or at the ever-receding margin of the growing city, such are the market rules.* Land acquisition is going on as the requirement of land for industrialization and urbanization is still high. *The port alone needs another 300 acres of land for shore disposal of silts.* We have categorized lands into swamp, agri-land as well as high land and homestead land, etc. The price is determined based on average of the reasonable market price of previous three years. Haldia Development Authority has notified some areas where the transfer, sale, and conversion of land is banned since the land would be acquired later. The farmers are opposed to acquisition, particularly the small and marginal peasants since they would lose their livelihood. On the other hand, *big landholders and ready to sell off as they would get rid of their share-croppers as well as make capital gains*. Sharecroppers are supposed to get Rs 15,510 per acre since the acquisition is a *fait accompli*; they want to increase the valuation of land by challenging the compensation at court after the acquisition. Evictees will get four to six decimal homestead lands. However, unlike earlier generations of evictees, they would not get any agricultural land since such land is no more there around the whole area. In this process, they loose their agricultural roots. *However, industrial jobs are not assured either. So it is not sure what will be fate of the new generation of displaced peasants*. (Emphasis added)

The discussion in this section reveals that though the genesis of Haldia port development induced displacement was several decades ago, its effects are long standing and widespread, and are being felt by several generations of the one family. Also notable from the interviews is the sense of bitterness, anger, and betrayal felt by the average port worker towards the state government, the respective labour hiring institutions, and the Haldia Port Development Authority for not adequately protecting their rights to land and labour and fair compensation, which, over the period of two decades or so, is being seen as woefully inadequate. The interviews also reveal the extent to which port development can influence wider social and economic developments in the port hinterland, which, in this case, includes valuable arable land and both fresh and sea waterways used for fisheries.

Conclusion

The chapter has outlined some features of the substantial investment being undertaken to improve India's maritime trade through its redevelopment of its ports. The case study of Kolkata port has suggested that there may be considerable costs associated with the process of port redevelopment and has emphasised the human costs of development-induced displacement. The rationalisation of workforce numbers, increased casualisation, and de-unionisation of dockworker labour has led to worsening conditions and a worrying lack of mechanisms to monitor safety. Moreover, the process of land acquisition and displacement of mostly peasant producers has been iniquitous and corrupt, with only few families financially benefiting from the windfall of hyper land inflation in Haldia and its surrounding localities. There are certainly difficulties associated with attracting greater private sector investment into the older parts of the port complex. While increased private sector involvement in India's ports may yield considerable benefits in terms of increased efficiency and throughput, a more thorough analysis of the costs and benefits should include aspects usually overlooked in the optimistic projections of policymakers. Significantly, displacement effects of port development are felt not merely at the site of the port, but in the surrounding towns, villages and the various waterways (the port hinterland) where land costs have spiralled, migrant labourers and other itinerant workers have arrived, where the waterways have become polluted or their access is restricted, and where public resources such as schools, roads and medical centres are under severe strain to cope.

References

Alvares, C.A. 1992. *Science, Development, and Violence: The Revolt against Modernity*. New Delhi and New York: Oxford University Press.

Bardhan, P. 1984. *The Political Economy of Development* in India Oxford: Basil Blackwell.

Baviskar, A. 2004. *In the Belly of the River: Tribal Conflicts over Development in the Narmada Valley*. New Delhi: Oxford University Press.

Brooks, Mary R. 2004. The governance structure of ports. *Review of Network Economics*, 3(2), 168–83.

Binder, S. and Tosic, J. 2005. Refugees as a particular form of transnational migrations and social transformations: Socioanthropological and gender aspects. *Current Sociology*, 53(4), 607–24.

Chimni, B.S. 2009. The birth of a 'discipline': From refugee to forced migration studies. *Journal of Refugee Studies*, 22(1), 11–29.

Colson, E. 2003. Forced migration and the anthropological response. *Journal of Refugee Studies*, 16(1), 1–18.

Corbridge, S. and J. Harriss. 2000 *Reinventing India: Liberalization, Hindu Nationalism and Popular Democracy* Cambridge: Polity Press.

Department of Economic Affairs. 2009 *Position Paper on the Ports Sector in India* Ministry of Finance, Government of India.

Dwivedi, R. 1999. Displacement, risks and resistance: Local perceptions and actions in the Sardar Sarovar. *Development and Change*, 30(1), 43–78.

Dwivedi, R. 2002. Models and methods in development-induced displacement. Review article, *Development and Change*, 33(4), 709–32.

Faure, A. 2008. Social norms for population displacements caused by large dams France, 20th century: The example of the Tignes and Serre-Ponçon dams in the Alps and the Aigle and Bort-les-Orgues dams in Haute-Dordogne. *La Revue de GéographieAlpine/Journal of Alpine Research*, 1, 29–44.

Ganguly-Scrase, R. & G. Vogl. 2008. 'Ethnographies of gendered displacement: women's experiences in South Asia under neo-liberal globalisation', *Women's Studies International Forum*, 31, (1), 1–15.

Graham, C. 1994. *Safety Nets, Politics, and the Poor: Transitions to Market Economies*. Washington, D.C.: The Brookings Institution Press.

International Labour Organization (ILO). 1996. *Social and labour problems caused by structural adjustment in the port industry TMPI/1996*. Geneva: ILO.

International Transport Workers' Federation (ITF). 2004. *Port Reform in a Global Context* [Online]. Available at http://www.itf.org.uk/english/education/pdfs/portreform/factsheet1.pdf [accessed: 20 August 2010].

Judge, P.S. 1997. Response to dams and displacement in two Indian states. *Asian Survey*, 37(9), 840–51.

Kidwai, A.H., 1989, Ports in a national system of ports and cities, in: Broeze, F. (ed.) *Brides of the Sea: Port Cities of Asia from the 16th –20th Centuries*, University of Hawaii Press, Honolulu, 207–22.

Parasuraman, S. 1996. Development projects, displacement and outcomes for displaced: Two case studies. *Economic and Political Weekly*, 31(24, 15 June), 1529–32.

Parasuraman, S. 1999. *The Development Dilemma: Displacement in India*. London: Macmillan.

Planning Commission. 2002. *Tenth Five Year Plan 2002–7*, New Delhi: Government of India.

Ray, A. 2009. Vanishing islands: Blame on KoPT. *The Times of India* [Online, 3 April]. Available at: http://articles.timesofindia.indiatimes.com/2009-04-03/kolkata/28047165_1_nayachar-global-warming-kolkata-and-haldia-docks [accessed: 9 November 2010].

Saha, S. 2009. Nandi shadow on Haldia Port future Cloud on site to dump silt. *The Telegraph* [Online, 9 September]. Available at: http://www.telegraphindia.com/1090917/jsp/nation/story_11505135.jsp [accessed: 14 November 2009].

Scudder, T. 2005. *The Future of Large Dams: Dealing with Social, Environmental, Institutional and Political Costs*. London: Earthscan.

Turnbull, P. 2001. *Contesting Globalization on the Waterfront* [Online]. Available at: http://depts.Washington.edu/pcls/turnbull.pdf [accessed: 10 November 2010].

United Nations Economic and Social Commission for Asia and the Pacific (UNESCAP). 2005. *Regional Shipping and Port Development Strategies (Container Traffic Forecast)* [Online: United Nations, New York]. Available at: http://www.unescap.org/ttdw/publications/tfs_pubs/pub_2398/pub_2398_fulltext.pdf [accessed: 10 November 2010].

World Bank. 1995. *India Port Sector Strategy Report*. Washington, D.C.: The World Bank.

Chapter 3
Youth on the Neoliberal Margins: Ethnography of Tourism-led Displacement in Kerala, South India

Belinda Green

Introduction

In the contemporary setting, the political economy of international tourism in Kerala, south India, is undergoing significant changes as a result of neoliberal policy and praxis. The tourist site of Kovalam in Kerala, a beachside tourist resort situated 16 kilometres from the state capital, Trivandrum, provides a disturbing exemplar of neoliberalism where privatisation, multinational investment, and the development of elite resorts and mass tourism has led to multiple forms of forced displacement and dislocation of local people. This chapter will provide an ethnographic account of the ways in which these processes are being met with and negotiated by a group of Scheduled Caste (SC) male youth who live and work in Kovalam. Rather than relying solely on two-dimensional constructions of powerless and marginalised victims, who are unable to participate in the neoliberal project, this chapter highlights the dynamic and complex response to such conditions by these young SC men. Through their articulation of a subcultural formation, locally referred to as the 'beach boys', they seek to engage with and utilise neoliberal forces for their own social and economic mobility.

This chapter will discuss the impact of neoliberal discourses on international tourist development in Kerala. More specifically, I will focus on the tourist site of Kovalam. Based on ethnographic data collected between 2003 and 2005, this chapter will explore the transformation of international tourism in Kovalam from its inception as a hippie haven in the 1960s to its current status as a globalised tourist site for mass tourist consumption. My aim is to highlight the ways in which state discourses pertaining to international tourist development in Kerala have become increasingly embellished by neoliberalism and the ways in which this has adversely affected the local population. Yet rather than reinforce ideas that local people are merely powerless victims to the former processes, the closing section of this chapter will detail the innovative response to neoliberal development by a group of young SC men who live and work in Kovalam.

According to Larner (2000: 5), neoliberalism 'denotes new forms of political-economic governance premised on the extension of market relationships' and that it 'is associated with the preference for a minimalist state. Markets are understood to be a better way of organizing economic activity because they are associated with competition, economic efficiency and choice' (Larner 2000: 5).

In the Indian setting, neoliberalism has become 'the new orthodoxy in development' (Scrase 2006: 2; also see Brohman 1995; Gosovic 2000). Chase-Dunn et al. (2000: 77) argue that the idea of globalisation is now synonymous with a hegemonic neoliberal ideology that 'celebrates the victory of capitalism over socialism and proclaims marketisation and privatisation as solutions to the world's problems'. According to Scrase (2006: 2), 'the resurrection and hegemony of market driven approaches identify state intervention as inefficient and counterproductive and thereby call for developing countries to privatise state owned enterprises, adopt a range of stabilisation measures to address balance of payment crises, and limit public expenditure'.

Tourism in Neoliberal Kerala

Although Kerala is not a powerful participant in the Indian economic liberalisation process, it is no stranger to global economic forces and flows of monies from abroad. In the context of migration flows, which have continued for decades, those having access to economic globalisation by being able to work abroad – in the Gulf and Arab world – have emerged as a highly consumerist and powerful force in the state (Kannan and Hari 2002: 201; Osella and Osella 2000: 123).

Yet with its continued reliance on agricultural exports in the face of an increasingly competitive and volatile global market, the Kerala government has been forced to re-evaluate its fiscal policy in an attempt to locate a place in the neoliberal context (Prakash 2004; Sreekumar and Parayil 2002: 529). One possible solution for a region which is well-endowed with natural resources is the development of international tourism.

Tourism offers the state precious foreign currency with relatively low levels of planning and investment. Although the promotion of tourism in Kerala has been actively promoted by the state since the early 1980s, in more recent decades, there has been a growing push towards privatisation and foreign and domestic investment (Netto 1999). This, in case of Kovalam, has been fraught with political, social, economic, and ecological side effects.

With increasing foreign and domestic investment, coupled with corruption and government neglect, there has been privatisation, multinational investment, and the development of elite resorts and mass tourism in the region, which has, in turn, led to the dislocation of local people who are also suffering under the weight of rising land and consumer prices within the area. Additionally, overcrowding, electricity and water shortages, and an overall lack of infrastructure, with the exception of a heavy-handed police presence has left many locals economically disadvantaged, dispossessed, and further dislocated.

The Tourist Site of Kovalam

Kovalam village falls in the Taluk of Neyyattinkara and the beach destination is located in both the Panchayat of Vizhinjam and the Panchayat of Venganoor (DESK 2001: 2; Sridhar and Nair 2004: 27). Kovalam junction is situated between the beach side area and the nearby Vizhinjam township. The two coves which make up the tourist area are approximately 2 kilometres from the junction. It is this beach area that this chapter will refer to as Kovalam.

As outlined by Jacob (1998: 42), prior to the emergence of international tourism in the area, Kovalam was 'a coconut village'. The word *kovalam* literarily means coconut grove. Coir products, copra oil, toddy, and other derivatives were all part of a local-based coconut industry owned and administered by an Izhavas elite. These products were produced by a local proletarian class, which largely comprised of Izhavas and a collection of Dalit groups.

Ever since the mid-1960s, this fishing- and agricultural-based village society has slowly emerged as one of the leading international tourist resorts in Kerala. Kovalam is particularly popular for its picturesque coast, calm waters, and beachside activities. During this early period, there was no organised accommodation for tourists in Kovalam nor was there any structured or regulated participation by locals. So called 'drifter' and 'explorer'[1] tourists, often referred to as 'hippies', were the first to venture to the area for tourist purposes, generally housing themselves by camping on the beach and surrounding areas.

After the interest displayed by adventurers and drifters from the Global North, several Izhavas land-owning classes began renting-out rooms in their homes to these tourists. This type of accommodation is described as a 'homestay'. From the money they accrued from this increasingly lucrative trade, some families extended their homes to further accommodate tourists. Some even managed to create small hotels from these rather unassuming beginnings.

The Indian Central and Kerala State Departments of Tourism first took control of parts of the surrounding beach area in 1966. With the exclusion of an area which was reserved for the construction of a hotel by the Kerala Tourism Development Corporation (KTDC), the state government handed over part of the land to the central Indian government. In turn, the India Tourism Development Corporation (ITDC) initiated the first state-run accommodation in Kovalam. The Ashok Hotel was built by the ITDC in 1973. This hotel/resort consisted of 65 acres of land, which included the 'Halycon Castle' originally built for the Maharaja of Travancore. Following closely behind, Hotel Samudra, owned by the state government, located on the northern cliff adjacent to the Ashok was established (Sridhar and Nair 2004: 18). In 1975, the Kovalam-Vizhinjam Development Authority was also founded.

This marked the beginnings of the central and state governments' ascending interest and investment in international tourism as a form of development and foreign currency in Kovalam. The state government was put in charge of the daily

1 These terms were first defined by Erik Cohen (1979).

running and administration of the Ashok Resort. This development also gave rise to the emergence of a more elite type of tourism and tourists travelling to Kovalam. During this time, the resort marked the transference of tourism in Kovalam from being a local-based industry to a state-sponsored enterprise. It also marked the beginning of the state's exploitation of the Kovalam people and the land.

According to Jacob (1998: 58) and primary sources, the 65 acres which made up the hotel grounds was purchased from local Muslim fisher people by the government under the banner of development for less than 350 INR for each cent[2] of land. As an added incentive, the people were promised jobs in the new resort as a means of encouraging them to abandon their agricultural enterprise on their land, which supplemented their vocation as fisher people. Although some were given jobs, these promises made by the central and state governments have been overwhelmingly unfulfilled. According to Sridhar and Nair (2004: 19) and Jacob (1998), the construction of the ITDC marked the beginning of a series of economic, social, and physical forms of displacement for this particular fisher community.

For those people who did manage to acquire employment from the resort, they were placed in roles which were by and large unskilled and at the lower end of the labour force. For instance, some were allotted jobs in the laundry department, as security personnel, or as sweepers and gardeners. Many of these jobs were also on a seasonal and casual basis. Alongside this, Jacob (1998: 58) alleged that the ITDC and the KTDC 'auction off their vacancies to the highest bidders' and are noted for corruption which has increasingly led to people from outside the area being recruited for positions in government-run facilities.

During the earlier stages of tourism in Kovalam, most of the restaurants located on the beach were no more than makeshift beach huts, made from coconut leaves with thatched rooves and sand floors. Given that the overwhelming majority of the beachfront land was owned by an Izhavas family, most of the restaurants were either owned directly by the family or a relative with the land being rented by local entrepreneurs on an annual basis. The majority of tourists who travelled to Kovalam during this phase were backpackers and other independent tourists from the Global North.

At that time, the accompanying rock formations segregating the two main beach coves was also utilised by local traders and entrepreneurs. As several informants explained, during this period there was a lot more involvement between tourists and locals especially on the beach itself.

For example, Rama, his mother, and brother established a small local restaurant on the rock area alongside other entrepreneurial fisher families who had set up makeshift mixed businesses and tea shops. Not only did these shops cater to the fishermen and others locals' needs, they also serviced the tourists. According to informants, during this stage of tourism in Kovalam, there was much more freedom to interact with the tourists. As Rama explains:

2 In South India, 'cent' is a term for land measurement. One cent of land is equivalent to 1/100th of an acre or 435.6 square feet

> We used to speak with the tourists on the beach with no one disturbing us. No police, no tourist police asking us what we are doing talking with a *Madamma* [foreign women] or *Sahib* [foreign man]. I could take the tourists out on my boat for fishing or snorkelling. Today it's not like that anymore, ever since the charter flights coming.

Although a number of informants cited the arrival of the charter flights as the beginning of the end of local involvement in the tourist industry, by the late 1980s and early 1990s, the government and to a lesser extent private foreign and domestic entrepreneurs had already set about establishing a more exclusive and state-monopolised form of tourism, which inevitably stood to further ostracise and displace local people from their environment. During the monsoon of 1993, this prosperous enterprise by Rama and his family was demolished by the government alongside 66 other 'illegal constructions' (Sridhar and Nair 2004: 18) including houses, hotels, and shops on the rock area on the beach.

The introduction of organised mass tourism in 1995 changed the face of tourism in Kovalam forever. Not only did this change the composition of tourists who ventured here, it also increased the level of competition for local traders and entrepreneurs. This competition included a rise in the amount of Indian nationals and foreign syndicates including international tour operators and charter groups venturing to the area to profit from the tourist trade. This was further exacerbated by a surge of local development by smaller entrepreneurs and local families who anticipated that the charter groups would provide an overall increase in tourist numbers and expenditure. Alongside these changes to the composition of tourists and the tourist economy in Kovalam, surveillance by the state and local authorities also increased.

During this period, the Kerala government eradicated all small, locally owned businesses on the beach, and designated the land as government territory, while at the same time organised particular infrastructures to ensure Kovalam was kept under closer surveillance and control by state authorities. Some of these measures introduced by the state included a life guard service, a tourist police force, and a lit footpath running the full length of the 'Lighthouse beach' alongside an intensification of local police control over drug use, prostitution, and unlicensed alcohol trade in restaurants. Yet, as Jacob (1998: 26) notes, bribery and corruption continue to dictate the high demand for police postings in the area, with local authorities making a considerable 'bonus' in their dealings with local traders and restaurateurs involved in selling illicit substances.

Organised mass tourism also changed the temporal boundaries of tourist activity in Kovalam. Prior to this transformation, independent tourists visited Kovalam all year round with higher levels of tourists from the Global North during the dry season between October and April. However, with the introduction of mass tourism, the tourist season[3] and tourist population have somewhat diminished.

3 'Season' refers to the peak tourist time in Kovalam which is between November up to and including April.

The tourist economy is now centred on a series of European-based charter flights arriving every fortnight for a limited period of time.

These charter flights consist of 200–250 British and northern European tourists who venture to Kovalam on a prepaid package tour from a tour operator. These travel companies are offshore organisations predominantly from Europe, who have small offices based on the ground, acting as satellite administrators for the larger infrastructure oversees. This type of tourism has negative consequences for local businesses and informal traders. There are only a select number of hotels that accommodate these tourists in Kovalam. Most of these hotels have connecting restaurants and other facilities included in their establishments. Some of the packages include meal tickets and prepaid transportation and sightseeing. The tourists are encouraged by tour operators and hotel staff to remain within the confines of the hotels and chartered-based activities. They are also actively discouraged from venturing outside with stories that local traders are capable of theft, cheating, and lying.

These events have had serious ramifications for Kovalam's local traders and their engagement with the tourists themselves. Cohen's (1979) characterisation of the 'drifter' and the 'explorer' is that they often stay longer than other tourist types while developing a greater awareness of 'the hosts and their culture' (Burns 1999: 44). On the other hand, 'mass tourists', identified as those who partake in organised, prepaid packaged tourist destinations, have 'minimal engagement with the locals and relatively little inter-cultural contact' (Burns 1999: 44).

Informants who had worked in tourist-related jobs in Kovalam since the 1980s confirmed Cohen's ideas. According to Sanjesh, a local hotelier – who, during the mid-1980s and before, worked as a tour guide and waiter in a number of establishments on the beach – the introduction of mass tourism into Kovalam had largely been a negative experience for him and other local businesses. He says:

> Before the charter groups came, we had a lot more freedom to mix with the tourist. There were no tourist police and no charter companies telling the tourists not to mix with the locals. Also the travellers don't like to come and stay in Kovalam anymore. They go to Varkala[4] instead. Too many people trying to sell them things and too many people looking at them on the beach. The charter groups don't come to the beach for swimming. Most of them stay in the hotel and sit by the pool. They don't want to buy things from the people and they are told by the hotels not to speak to the people. Kovalam is not the same anymore. Too many people trying to cheat the people and not enough tourists to make good business.

Kovalam's metamorphism over the past 50 years posed many questions. Given that I had visited Kovalam for the first time in 1994, from what was once was a series of thatched roof beach hut restaurants with a few hotels dotting the

4 Varkala, located in Kollam, Kerala, is another international tourist site.

periphery of the main beachfront known as Lighthouse Beach, Kovalam has much transformed into a globalised tourist resort today. Complete with an extensive range of tourist accommodation, restaurants that had mostly abandoned the thatch roof beach huts for concrete and makeshift materials alongside a multitude of tourist services including Internet cafés, Ayurvedic treatment centres, handicraft shops, tailors, jewellers, tour operators, auto rickshaws, and taxi services, it was no longer a place where the backpacker could find orientalist solace in the picturesque landscape that places like Kovalam offer.

Therefore, in more recent times, the tourist resort of Kovalam has become increasingly embellished by neoliberal forces where private entrepreneurial investment, multinational forces, and a movement away from state-sponsored enterprise has had an adverse effect on the local ecology and its people.

Large-scale hoteliers require large tracts of land for their ambitions to transform Kovalam into an exclusive resort style enclave. This has resulted in corrupt state practices with the government re-zoning public land for a quick profit. For example The Taj Hotel Group are the most recent contingent to build a resort called The Taj Green Grove Resort at Samudra Beach. Not only has this establishment further alienated the local people from the land market, it has also dispossessed a number of fisher people of their livelihood on the nearby beach. With the beach designated as government-owned property, the Taj Group has managed to acquire part of the beach for their customers' exclusive use. This decision caused upheaval amongst the local fisherman, given that they and previous fisher people had been working in this area for several successive generations. A number of protests were organised by local unions against the government sale yet to no avail. This scenario is not an isolated case.

There are other parts of the coastal areas such as the small inlet in front of the state-government run Hotel Samudra which is 'off limits' to anyone but fishermen for their livelihood and of course the tourists for sunbathing and swimming. All other locals are prohibited from entering this area. It is policed by the hotel's security guard, the beach chair, and umbrella boys, all hired by the hotel alongside the local lifeguards. A part of the Ashoka beach cove has also been partitioned by the Ashok five-star resort for the sole use of its guests. Even the fishermen are prohibited from entering this area. Security men are strategically positioned to ensure that the tourists are 'protected' from locals.

The presence of surveillance over the local population has also been on the increase over the last decade. At the same time, as the charter groups moved into the area and established their pre-packaged holidays, a tourist police service was also established. These policemen are specifically focused on protecting the tourists and actively deterring the locals from interacting with international visitors. They, along with the local police, patrol the beach and surrounding areas, adorning a blue uniform compared to the police's military green style.

The introduction of a state-sponsored lifeguard service also provides additional surveillance of the beach area. This service was also put into to place to 'protect' the tourists, which translates into keeping local peoples at a distance from the foreigners. Informants who had been able to create a lot of business from creating interpersonal rapport with the tourists through regular contact on the beachfront now find themselves being pushed out of the trade altogether. Rama explains, 'we aren't even allowed on our own beach anymore in season time'.

Aside from the fishermen, it is very difficult during season time for local men in particular to venture onto the beach unless they are specifically involved in the tourist trade, either working at one of the restaurants or hotels or selling goods in shops or working as hawkers. These heightened levels of surveillance have intensified social and economic disparities between local hosts and foreign mass tourists. The mass foreign tourists are deliberately kept at a distance through the assistance of the aforementioned intermediaries.

Issues of displacement, overcrowding, environmental degradation, increasing inequalities in the distribution of wealth and living standards as a result of the increase in commercialisation, and internationalisation of the Kovalam tourist industry has furthered the marginalisation of local peoples.[5] Although there continues to be a local elite of Izhavas and to a lesser extent a Muslim elite, these former groups comprise less than 5 per cent of the local population. As noted by Jacob (1998), Sridhar and Nair (2004), and Vijaykumar (1993), the majority of the local population, that is, the fisher people, Other Backward Classes (OBC), and SC communities in the immediate and surrounding areas remain subjected to impoverished living conditions.

Over the past decade, there has been a significant amount of construction and development in this area. As a result, parts of Kovalam had suffered serious environmental degradation. Due to people taking sand from the beaches for construction purposes, the two main beach coves were half their size since my first visit in 1994. The wide stretch of sands, which separated out the beach from the eateries and shops, had all but disappeared. Nowadays, there is no delineation between the shore and the swelling expanse of beachfront construction. Those at the beach are in full view of those eating at the restaurants or the constant flow of pedestrians venturing along the newly constructed footpath.

The beach enthusiasts are also exposed to the multitude of hawkers who can often outnumber the actual tourists on the beach. From what was a few fruit ladies selling local tropical fruits to tourists on the sand, the sunbathers are now being offered everything from sandalwood carvings and beach clothes to flashing gadgets of no real purpose or appeal. Calls of 'Fruit Salad?' 'Pineapple?' 'Banana?' 'Look madman nice sarong … very nice', 'Where are you from?' 'What is your good name?' are a part of the beach experience for the tourists.

Given that the majority of international mass tourists had ventured to Kovalam for as a recreational getaway, many expressed discontent with the hawkers' persistence. As one British tourist, Ed, recounts:

5 See Sridhar and Nair (2004: 33–34) for further information on environmental degradation including over fertilisation of soil.

> I only get a few weeks holiday a year and I like to come to India because it is cheap and the weather is good. But the people here are too pushy and they don't take no for an answer. All my wife and I want to do is relax in the sun, enjoy the beach, and do some sunbathing. Half of the stuff they [the hawkers] are selling is crap that falls to bits before you even get it home! We're not interested in buying it but they just keep asking. We end up having to be rude or moving off the beach and sunbathing by the pool.

These ideas were also confirmed by Sridhair and Nair's 2004 study (p. 31). The constricted spatial boundaries have clearly increased tensions between the tourist and host communities. In reality, tourists are the least affected by the problem of overcrowding in Kovalam. With a local population of just over 25,000 people in the Kovalam ward, Sridhar and Nair (2004: 29) estimate that any one day during the peak tourist season, the population can triple in size.

The rising rate of land prices both in the rental and purchasing markets has been a major source of contention and anxiety for local communities over the past 15 years in particular. With the increasing folds of people from other parts of the state, India and international agents, including small-scale merchants, foreign expatriates, domestic and foreign entrepreneurs, and multinational syndicates, the value of land has skyrocketed. Even in the early stages of charter tourism, Jacob (1998: 55) had found that there had been a '25 fold increase' for the price of Kovalam land since the early 1980s. Alongside this the price for renting beach front land in 1996 had doubled in just one year (Jacob 1998: 26).

From 25,000INR for a 10 ft by 5 ft piece of land during the tourist season in 1997, today's price can be anything from 1–2 lakhs (100,000INR–200,000INR). Selling one's land no matter how small the plot has become an attractive option for many poor Kovalam families. The disputes over land settlements between family members, the provisions of pathways for people's newly built or existing homes which may intersect with other people's lands, is a continuing source of conflict and tension for local families.

Therefore, the overwhelming expense of the land in the area has not only edged out the local population from participating in the market, it has also excluded any smaller investors. Multinational hotel chains and larger domestic and foreign syndicates constitute the buyers of what little land remains left. On top of this, these larger scale syndicates do not employ local people. As previously mentioned at best local people were once hired for the lower end jobs. Today these positions are filled by people from outside of the area which may include the cheap pool of labour coming from other parts of India who are not supported by any local unions. According to Sridhar and Nair (2004: 30), 40 per cent of hotel ownership is from people outside of the area who bring in their own staff. The researchers also found that even hotels owned by local people were also staffed by people outside of the area (Sridhar and Nair 2004: 30).

The continuing and longstanding neglect of state-run enterprises towards it people all over India is further exacerbated for many communities under the harsh light of neoliberal development. Kovalam is no exception. Water and electricity shortages and ineffective waste management have all become a regular part of daily life for lower-class groups in Kovalam. With the exception of the tourists and those wealthy enough to have generators, these conditions have become increasingly worse. Sridhar and Nair (2004: 31) have also found that the area seriously lacks adequate infrastructure, including 'education, communication, transportation, recreation, health, etc'.

With its crescent shaped coastline, fishing, granite cliffs, coconut trees, and calm ocean coves, it is perhaps not surprising that Kovalam has emerged as an internationally renowned tourist resort. However, like others who have written about its tourist industry (see Jacob 1998; Sridhar and Nair 2004; Vijaykumar 1993), the exploitation and subsequent neglect of the local ecosystem and its inhabitants as a result of its increasing commercialisation is overwhelming. Nevertheless, Kovalam continues to expand on its touristic contours with an increasing consciousness or projection at least of a more eco-friendly approach to tourism by both government and non-government forces.

On the social and political front, Jacob (1998), Sridhair and Nair (2005), and Henderson and Weisgrau (2007) argue that tourism is bad for local people who are increasingly being pushed aside in the tourist industry in a place like Kovalam and Rajasthan as a result of neoliberal forces of privatisation and multinationals. State support for such ventures combined with a lack of infrastructural and alternative employment begins to mark out the hardships faced by local people in tourist sites and spaces in India. Although tourist sites like Kovalam previously offered relatively unskilled local people an opportunity to make money servicing the independent tourists, in the contemporary setting, these chances are becoming less likely. This is a result of neoliberal tendencies where state participation is now characterised by increased surveillance and control over local populations in tourist settings to facilitate private enterprise and inter and intra-state investment in the development of international tourism. As a result groups of local people in Kovalam are being forced to adopt underground and other illicit labour strategies in an attempt to etch out a living wage for themselves including the SC young men as discussed in this chapter.

The situation in Kovalam is not an isolated case. Overall, the process of economic liberalisation in India has resulted in further socio-economic marginalisation for many groups as well as furthering the polarisation of class relations. In the contemporary context, job loss, underemployment, casualisation, and feminisation of the lower strata of the labour force have been particularly felt by tenant farmers, women, labourers, artisans, and those workers from the SC and Scheduled Tribes (ST)[6] categories (Acharya 1995; Afshar and Barrientos 1999; Appadurai and Breckenridge 1995; Arora 1999; Bannerjee 2002; Basu 1996; Bhattacharya 1999; Chakraborty 2004; Channa 2004; Colloredo-Mansfield

6 The terms SC and ST refer to peoples who were formerly known as outcastes and untouchables in the Hindu pantheon

2002; De Neve 2008; Dewan 1999; Ganguly-Scrase 2001, 2007; Omvedt 1997; Ramachandran and Swaminathan 2002; Scrase et al. 2003; Sudarshan and Mukhopadhyay 2003). Even though it is clear that such groups are facing further impoverishment as a result of neoliberal development, these findings reinforce that marginalised people are mere victims of the former processes, exploited and powerless to alter their socio-economic position, or engage with the processes of economic and cultural globalisation.

The last section of this chapter seeks to disrupt these ideas by exploring the ways in which a group of marginalised young men are responding to and engaging with the forces of neoliberalism and globalisation in the Kovalam setting. As noted by Appadurai (1990: 9) in his landmark work on cultural globalisation in Asia, the politics of identity are being increasingly influenced by what he describes as 'possible lives'. The idea of possible lives is a result of people's lives and frame of reference being expanded beyond the everyday through 'education, travel and the consumption of mass media' (Liechty 1995: 167).

One of the most widespread and powerful forces to come out of economic liberalisation in India is the proliferation of global media forms (Appadurai 1996; Butcher 2003; Ninan 1995; Scrase 2002). For Giddens (1990), Sklair (1991), and Waters (1995), the proliferation of technological advancement through time–space compression and new media forms has significantly impacted on identity formation. Accordingly, the resultant weaving and disjuncture of the global with the local has created an 'identity crisis' for social groups (see Castells 1997; Giddens 1991), which is reconciled by 'forms of group self-invention in lifestyle and consumption practices' (Nilan and Feixa 2006: 6).

All over Asia, young people in particular find themselves in what Liechty (1995: 190) defines as an 'in-between space – between expectations and reality; between past and future; between village and an external modern metropolis; between child and adulthood; between high and low class; between education and meaningful employment' (see also Nilan and Feixa 2006: 2).One way to manage such anxieties for youth populations is to participate in peer groups. As argued by Liechty (1995: 190) 'peer groups allow young people to abandon themselves in the material existence, consciously avoiding the future by living for each other in the present' (see also Lukose 2005; Rogers 2005, 2008). These group identities are based around 'conformity to group dictated standards of taste in dress' (Liechty 1995: 190) and food, music, drugs, and so on. Peer groups serve as a way to manage the anxiety with lived realities versus the possible. As noted by Nilan and Feixa youth cultures' reflexive engagement and process of cultural hybridity is also contingent and shaped by their habitus which includes a multiple array of variables including 'income, religion, language, class, gender and ethnicity' (2006: 8).

The young men in this chapter utilise forms of popular culture and global artifacts within their habitus to articulate a hybridised subcultural identity locally referred to as the 'beach boy'. The beach boys have created this cultural hybridisation through a complex mediation of local and global forms of consumption and production. The young men's engagement with international tourism and transnational sexual/

romantic relations with tourists from the Global North has created a heterogeneous and complex identity. Appropriating their subcultural performativity for tourist consumption, the young men seek an alternate strategy for economic and social mobility in the contemporary neoliberal period.

The label 'beach boys' is a familiar term in sex/romance tourism studies[7] (see Brown 1992; de Albuquerque 1998; Herold et al. 2001). This describes local male sex workers who service women tourists from the Global North in tropical 'out of way' tourist places.[8] As a result of their underclass status and marginalisation in the local setting, the beach boys seek to raise their socio-economic status through transnational sexual intimacies with women tourists from the Global North holidaying in Kovalam.

Third World male hosts seeking economic mobility through sex/romance work with women tourists from the Global North have been investigated by a cross section of scholars in anthropology, sociology, and human geography (Bowman 1989; Brown 1992; Dahles and Bras 1999; de Albuquerque 1998; Henderson and Weisgrau 2007; Herold et al. 2001; Jennaway 2002; Lette 1996; Malam 2003, 2006; Meisch 1995; Pruitt and LaFont 1995; Wolf 1993).

The informal tourist economy, although particular in its local conventions, is not unique to Kovalam, nor is the presence of young groups of male youth who are often on the front line of such economic activity (Beazley and Chakraborty 2008; Dahles and Bras 1999; Herold et al. 2001). Compared to the alternatives found in the Kovalam context which might include head-loading labour or chopping granite on the side of the road for just under an AUS$1 a day, making money through tourism is a far more attractive option to these young men. This scenario is particularly evident in the neoliberal context where economic rationalism, consumption, and commoditisation of self are the driving forces behind social and economic status.

The beach boys in Kovalam are made up groups of SC and OBC[9] male youth aged between 18–35 who live and work in tourist resorts in Kerala like Kovalam.

7 Pruitt and LaFont (1995) first coined the term 'Romance Tourism' to describe encounters between the women tourists and host men in the Caribbean. The writers found that these encounters often involved a period of courtship with an emphasis on romance extending over a period of time as opposed to a one-off sexual encounter. An initial and open exchange of money for sex was often replaced by an 'indirect' and convoluted exchange of love, romance, sex, alcoholic beverages, food, money, gifts, outings, etc., which may have lasted the entire duration of the tourist's stay in the host's country. Informants also claimed to be emotionally involved with one another, with several of these encounters extending into long-term relationships.

8 The use 'out of way places' was first adopted by Anna Tsing (1993) as an alternative to value laden terms such as 'Third World', 'Developing World', and 'South'.

9 Further to identifying their former status as untouchables, the categories of SC, OBC, and ST are legislative titles introduced by Ambedkar's Indian Constitution in the early 1950s. This awarded to former untouchable groups in India, which is centered on the praxis of positive discrimination through a nation, wide reservation quota system.

They are from a variety of religions, *jatis*,[10] and family backgrounds, yet share a similar class position and low socio-economic status. Local jatis or subcastes which make up the majority of the beach boys in Kovalam include SC Pulayan male youth and to a lesser extent OBCs including Christian Mukkuvar, Muslim Marakar, and Izhavas male youth. Beach boys can also be found in other localities including Varkala in nearby Kollam and are also located in other parts of India's west coast including Goa and Karnataka. There are close to 100 beach boys who inhabit the tourist sites of Varkala and Kovalam in Kerala.

This chapter focuses on 25 SC beach boys who work in the informal sector of the tourist economy in Kovalam. The findings of the research revealed the regularity of underemployment or outright unemployment in the contemporary neoliberal setting for SC people in the area. The overwhelming majority of informants had no work during the long monsoon months outside of the tourist season. Occasionally, some were able to find work as labourers or as mason assistants during that time. However, this was a rare occurrence. Just under half of the informants did not have any permanent jobs, even during the tourist season, however, many of them worked on an ad hoc basis for tourists in terms of purchasing drugs, arranging taxis, and working as guides. The half that did gain some employment during the tourist season was situated mainly in the informal and lower end of the tourist market, with the other half of the informants were self-employed as beach chair and umbrella boys, fishermen, or auto rickshaw drivers; only two were employed in the formal sector with one of them being employed overseas as a result of marrying a foreign woman and the other having a government job as a beach lifeguard in Kovalam. Those who were employed in the informal tourist sector and did not work for themselves were waiters, kitchen hands, or night watchmen for small hotels.

None of these young men had been able to pursue their education beyond secondary school. Just over half of them completed secondary schooling with the majority making it only as far as eighth standard. Out of the 25 informants, only three had completed 12th standard. The average age of the beach boys was 25. Out of the 25 beach boys and their families in the study living in the Kovalam and surrounding areas, the maximum amount of land owned by per family was 10 cents,[11] with three families owning as little as 3 cents of land. The average size of land owned by the families in the study living in Kovalam was just under 8 cents.

The earnings of the beach boys varied and were subject to the time of year, the availability, and type of work undertaken. The average wage for a job as a waiter working 12 hours six days per week on the beach was 1,200INR (AUS$40) per month plus tips. For those who were self-employed and worked in the tourist industry, during the season time, monies earned could vary anywhere from 100 INR up to 3,000 INR (AUS$100) per day. Although some beach boys who have an asset including beach chairs and umbrellas or an auto rickshaw and belong to a

10 'Jati' literally means blood group, that is, belonging to one's subcaste group.

11 A cent is equivalent to one-hundredth of an acre.

union can make considerably more money than their caste counterparts involved in fishing, agricultural, or laboring work, the money accrued during the tourist season is the beach boys' main source of financial support throughout the year, which includes the duration of the long monsoon months.

In more recent times, the impact of globalisation and neoliberalism has resulted in ideals of progress and development being reinforced through mobility and Non-resident Indian (NRI) status. For the beach boys, these ideals of mobility and NRI status are located in the hope and desire to marry a European woman and migrate to Europe in order to increase wealth and consumption for present and future self and kin. The young men also aim to accumulate large sums of money or receive valuable assets such as a house, motor vehicles, or a business venture vis-à-vis global north tourist 'friends' to improve theirs and their families' livelihood within Kovalam.

Therefore, the beach boy phenomenon is primarily centred on the hope of economic and social mobility through what they themselves refer to as *catching a tourist*. In order to do so, the young men find ways to appeal to the foreign tourists. This includes adopting particular forms of dress, behaviour, and interpersonal skills and vocations including sexual advances. Although the purpose of the beach boys is to catch a tourist, they simultaneously create a form of 'subcultural capital'[12] between one another to rise above their lowly socio-economic position as lowly educated SC young men.

Through their consumption of global and local forms of mass media, forms of dress, language, and engagement with international tourism in Kovalam, the beach boys attempt to re-posit their marginalised status through a conscious commoditisation of self for tourist consumption. The young men's aspirations to leave India are in part based on their belief that India provides very little opportunity to them as SC male youth with limited education and low socio-economic status in the contemporary neoliberal context. The following quote by Rama epitomises the beach boys' feelings of marginalisation and desire to escape the power relations in Kovalam:

> Sometimes I hate the fucking Kovalam. Everybody trying to make the money from the tourists, nobody caring about each other. You can't trust the people! People fighting over the land, over the tourists' money, too much drinking and fighting with the people here. The Izhavas have the power and money in

---

12 This term was first coined by Sarah Thornton (1995) from the Birmingham School of Cultural Studies in her study on club subculture in the UK. Thornton utilises Pierre Bourdieu's work on capital with particular reference to cultural capital to explain the ways in which subcultures develop particularised forms of cultural capital loosely based on group membership within the group itself. Unlike Bourdieu's reference to forms of cultural capital which prevail in the wider society, Thornton's term 'subcultural capital' refers to the way in which subculture develop their own forms of cultural capital, that is, 'subcultural capital'.

> Kovalam. If I getting the chance one day, I want to leave fucking Kovalam. This life is too hard for me!

Migrating overseas for the beach boys has become 'an essential means of accumulation of economic and symbolic resources which can be mobilised as part of a wider project of social mobility' (Gardner and Osella 2003: viii). Migration to the Global North represents a number of possibilities and aspirations for the beach boys. Ideas of a particular lifestyle, increased levels of consumption, and overall economic mobility for family members, escaping their place of marginalisation and poverty, while acquiring the desired transnational identity of NRI are all apart of the beach boy *hero dream*.

The beach boy subculture seeks to increase levels of mobility and consumption for the young men, but also attempts to etch out a place for their selves within the evolving and changing nature of the neoliberal Indian state. As noted by Stuart Hall (1988: 24), the processes of mobility and globalisation create 'greater fragmentation and pluralism, weakening of older collective solidarities and block identities and the emergence of new identities associated with greater work flexibility [alongside] the maximisation of choice through personal consumption'. These processes are further enhanced by other forms of globalisation including the collapsing of time and space through increased technological advancement, whereby the young men can continue the contact with foreign tourists even after the tourists' departure. The evolution of Kovalam as an international globalised tourist site has in part facilitated such processes for groups such as the beach boys.

Conclusion

As outlined in this chapter, the tourist site of Kovalam in Kerala sets the scene for understanding the ways in which some marginalised young people in South Asia are responding to contemporary processes of neoliberalism, transnational flows, and globalisation. The progression of international tourism in Kovalam provides an ideal backdrop. With increasing levels of private investment, corruption, and state surveillance, Kovalam has undergone a marked transition from its beginnings as a hippie haven to its present condition as a globalised international tourist resort.

Fernandes's (2004) work on the effects of neoliberalism on the urban development of Mumbai provides a similar scenario to the development of tourism in Kovalam. The writer argues that as a result of neoliberal discourses the Indian state has become complicit with what Fernandes labels as the 'the politics of forgetting'.

This idea refers to a political-discursive process in which specific marginalised social groups are rendered invisible within the dominant national political culture. Such dynamics unfold through the spatial reconfiguration of class inequalities. Both middle-class groups and the state engage in a politics of forgetting that displaces the poor and working classes from such spaces. The result is the production of an exclusionary form of cultural citizenship which is, in turn, contested by these marginalised socio-economic groups (Fernandes 2004: 2415).

The chapter highlights the problems of displacement, rising commodity and land prices, and an overall lack of infrastructure for local people including the beach boys and their families which has had an adverse effect on their living standards. Coupled with this the influx of global and local hotel syndicates and international tour operators supported on the ground by the state have seriously diminished local participation and informal trade within Kovalam and has, therefore, exacerbated issues of unemployment and underemployment for local traders. The progression of international tourism in Kovalam highlights the ways in which neoliberal tendencies have adversely affected the informal tourist economy. Yet, at the same time, this chapter has also highlighted the ways in which a group of marginalised young men are responding to these processes of change in complex and dynamic ways.

References

Acharya, S.K. 1995. Spatial implications of the new economic policy – Reflections on some issues from north east Indian with special reference to women. *Man and Development*, 17(1), 6–64.

Afshar, H. and Barrientos, S (eds). 1999. *Women, Globalization and Fragmentation in the Developing World*. London: Macmillan.

Appadurai, A. 1990. Disjuncture and difference in the global cultural economy. *Public Culture*, 2(2), 1–24.

Appadurai, A. 1996. *Modernity at Large: Cultural Dimensions of Globalization*. Minneapolis: University of Minnesota Press.

Appadurai, A. and Breckenridge, C (eds). 1995. *Consuming Modernity: Public Culture in a South Asian World*. Minneapolis: University of Minnesota Press.

Arora, D. 1999. Structural adjustment programs and gender concerns in India. *Journal of Contemporary Asia*, 29(1), 328–61.

Bannerjee, N. 2002. Between the devil and the deep sea: Shrinking options for women in contemporary India, in *The Violence of Development: The Politics of Identity, Gender and Social Inequalities in India*, edited by K. Kapadia. London: Zed Books, pp. 43–68.

Basu, R. 1996. New economic policies and social welfare programmes in India. *Social Action*, 46(3), 262–77.

Beazley, H. and Chakraborty, K. 2008. Cool consumption: Rasta, punk and Bollywood on the streets of Yogyakarta, Indonesia and Kolkata, India, in *Youth, Media and Culture in the Asia Pacific Region*, edited by U. Rodrigues and B. Smalls. Cambridge: Cambridge Scholars Publishing, pp. 195–214.

Bhattacharya, D. 1999. Political economy of reforms. *Economic and Political Weekly*, XXXIV(23), 1408–10.

Bowman, G. 1989. Fucking tourists: Sexual relations in tourism in Jerusalem's Old City. *Critique of Anthropology*, 9(2), 77–93.

Brohman, J. 1995. Universalism, eurocentrism, and ideological bias in development studies: From modernisation to neo-liberalism. *Third World Quarterly*, 16(1), 121–40.

Brown, N. 1992. Beach boys as cultural brokers in Bakua town, The Gambia. *Community Development Journal*, 27(4), 361–70.

Burns, P. 1999. *An Introduction to Tourism and Anthropology*. London: Routledge.

Butcher, M. 2003. *Transnational Television, Cultural Identity and Change: When STAR Came to India*. New Delhi: Sage Publications.

Castells, M. 1997. *The Power of Identity, The Information Age: Economy, Society and Culture*, Vol. 2. Oxford: Blackwell.

Chakraborty, D. 2004. Expansion of markets and women workers: Case study of garment manufacturing in India. *Economic and Political Weekly*, 39(45), 4910–16.

Channa, S.M. 2004. Globalization and modernity in India: A gendered critique. *Urban Anthropology and Studies of Cultural Systems and World Economic Development*, 33(1, Spring), 37–71.

Chase-Dunn, C., Kawano, Y., and Brewer, B. 2000. Trade globalisation since 1795: Waves of integration in the world-system. *American Sociological Review*, 65(1), 77–95.

Cohen, E. 1979. Rethinking the sociology of tourism. *Annals of Tourism Research*, 9(1), 18–35.

Colloredo-Mansfield, R. 2002. An ethnography of neo-liberalism. Understanding competition in artisan economies. *Current Anthropology*, 43(1), 113–38.

Dahles, H. and Bras, K. 1999. Entrepreneurs in romance: Tourism in Indonesia. *Annals of Tourism Research*, 26(2), 267–93.

de Albuquerque, K. 1998. Sex, beach boys and female tourists in the Caribbean. *Sexuality and Culture*, 2, 87–111.

De Neve, G. 2008. Global garment chains, local labour activism: New challenges to trade unionism and NGO activism in the Tiruppur industrial cluster, South India, in *Hidden Hands in the Market: Ethnographies of Fair Trade, Ethical Consumption and Corporate Social Responsibility*, edited by D. Wood, J. Pratt, P. Luetchford, G. De Neve, et al. *Research in Economic Anthropology*, 28, 213–40.

Department of Economics and Statistics Kerala (DESK). 2001. *Panchayat Level Statistics*. Thiruvathapuram: DESK.

Dewan, R. 1999. Gender implications of the ‘new’ economic policy: A conceptual overview. *Women Studies International Forum*, 22(4), 425–29.

Fernandes, L. 2004. The politics of forgetting: Class politics, state power and the restructuring of urban space in India. *Urban Studies*, 41(12), 2415–30.

Ganguly-Scrase, R. 2001. *Global Issues, Local Contexts: The Rabis Das of West Bengal*. London: Sangam Books.

Ganguly-Scrase, R. 2007. *Garment Workers in India in the Era of Liberalisation: Continuities and Changes*, Paper presented for International Convention of Asia Scholars, (ICAS), Kuala Lumpur, Malaysia, 2–5 August.

Gardner, K. and Osella, F. 2003. Introduction. *Contribution to Indian Sociology*, 37(1 and 2) (double special issue on 'Migration, Modernity and Social Transformations in South Asia' edited by F. Osella and K. Gardner; republished in 2004, Thousand Oakes, New Delhi: Sage Publications).

Giddens, A. 1990. *The Consequences of Modernity*. Cambridge: Polity Press.

Giddens, A. 1991. *Modernity and Self-Identity: Self and Society in the Late Modern Age*. Cambridge: Polity Press.

Gosovic, B. 2000. Global intellectual hegemony and the international development agenda. *International Social Science Journal*, 52, 447–56.

Hall, S. 1988. Brave New World, Marxism Today, October, 24–29

Henderson, C.E. and Weisgrau, M. (eds). 2007. *Raj Rhapsodies: Tourism, Heritage and the Seduction of History*. Burlington: Ashagate.

Herold, E., Garcia, R., and DeMoya, T. 2001. Female tourists and beach boys: Romance or sex tourism? *Annals of Tourism Research*, 28(4), 978–97.

Jacob, T.G. 1998. *Tales of Tourism in Kovalam*. Thiruvananthapuram: Odyssey.

Jennaway, M. 2002. *Sisters and Lovers: Women and Desire in Bali*. Lanham: Rowman & Littlefield.

Kannan, K.P. and Hari, K.S. 2002. Kerala's gulf connection: Remittances and their macroeconomic impact, in *Kerala's Gulf Connection: CDS Studies on International Labour Migration from Kerala State in India*, edited by K.C. Zachariah, K.P. Kanan, and S. Irudaya Rajan. Thiruvananthapuram: Centre for Development Studies, 199–230.

Larner, W. 2000. Neo-liberalism: Policy, ideology, governmentality. *Studies in Political Economy*, 63,Autumn 5–26.

Lette, H. 1996. Changing my thinking with a western woman: Javanese youths' constructions of masculinity. *Asia Pacific Viewpoint*, 37(2), 195–205.

Liechty, M. 1995. Media, markets and modernization: Youth identities and the experience of modernity in Kathmandu, Nepal, in *Youth Cultures: A Cross Cultural Perspective*, edited by V. Amit-Talia and H. Wulff. Princeton: Princeton University Press, pp 166–201

Lukose, R. 2005. Consuming globalization: Youth and gender in Kerala, India. *Journal of Social History*, 38(4, Special Issue: Globalization and Childhood), 915–35.

Malam, L. 2003. *Performing masculinity on the Thai beach scene*. Working Paper No. 1–15 Department of Human Geography, RSPAS, The Australian National University, Canberra.

Malam, L. 2006. Representing 'cross-cultural' relationships: Troubling essentialist visions of power and identity in a Thai tourist setting. *Acme: An International E-Journal for Critical Geographies*, 5(2), 279–99.

Meisch, L.A. 1995. Gringas and Otavalenos: Changing tourist relations. *Annals of Tourism Research*, 22(2), 441–62.

Netto, N. 1999. Tourism development in Kerala, in *Kerala's Economic Development: Performance and Problems in the Post-Liberalisation Period,,* edited by B.A. Prakash. New Delhi: Sage Publications, 269–92.

Nilan, P. and Feixa, C. (eds). 2006. *Global Youth? Hybrid Identities, Plural Worlds*. London: Routledge.

Ninan, S. 1995. *Through the Magic Window: Television and Change in India*. New Delhi: Penguin.

Omvedt, G. 1997. Rural women and the family in the era of liberalization: Indian in comparative Asian perspective. *Bulletin of Concerned Asian Scholars*, 29, 4.

Osella C. and Osella, F. 2000. *Social Mobility in Kerala: Modernity and Identity in Conflict*. London: Pluto Press.

Prakash, B.A. 2004. Economic backwardness and economic reforms in Kerala, in *Kerala's Economic Development: Performance and Problems in the Post-Liberalisation Period*, 2nd edition, edited by B.A. Prakash. New Delhi: Sage Publications, 32–58.

Pruitt, D. and LaFont, S. 1995. For love and money: Romance tourism in Jamaica. *Annals of Tourism Research*, 22(2), 422–40.

Ramachandran, V.K. and Swaminathan, M. 2002. Rural banking and landless labour household: Institutional reform and rural credit markets in India. *Journal of Agrarian Change*, 2(4), 502–44.

Rogers, M.C. 2005. *Tamil Youth: The Performance of Hierarchical Masculinities: an Anthropological Study of Youth Groups in Chennai, Tamil Nadu South India*, Unpublished DPhil thesis, Department of Anthropology, University of Sussex, Sussex.

Rogers, M.C. 2008, Modernity, 'authenticity', and ambivalence: Subaltern masculinities on a south India college campus. *Journal of the Royal Anthropological Institute*, 14, 1 79–95.

Scrase, T. 2002. Television, the middle classes and the transformation of cultural identities in West Bengal, India. *Gazette: The International Journal for Communication Studies*, 64(4), 323–42.

Scrase, T. 2006. The 'New' Middle Class in India: A re-assessment. 16th Biennial Conference of the Asian Studies Association of Australia, Wollongong, 26–29 June.

Scrase, T., Holden, T.J.M, and Baum, S. (eds). 2003. *Globalization, Culture and Inequality in Asia*. Melbourne: Trans Pacific Press.

Sklair, L. 1991. *Sociology of the Global System*. Brighton: Harvester-Wheatsheaf.

Sreekumar, T.T. and Parayil, G. 2002. Contentions and contradictions of tourism as development option: the case of Kerala, India. *Third World Quarterly*, 23(3), 529–48.

Sridhar, R. and Nair, S.K. 2004. Zero waste Kovalam and employment opportunities. *Kerala Research Programme on Local Level Development*. Thiruvananthapuram: Centre for Development Studies.

Sudarshan, R.M. and Mukhopadhyay, S. (eds). 2003. *Tracking Gender Equity under Economic Reforms: Continuity and Change in South Asia*. New Delhi: Kali for Women.

Thornton, S. 1995. *Club Culture: Music, Media and Subcultural Capital*. Cambridge: Polity Press.

Tsing, A.L. 1993. *In the Realm of the Diamond Queen: Marginality in an Out-of-the-way Place*. Princeton: Princeton University Press.

Vijaykumar, B. 1993. *Impact of Tourist Development – A Pilot Study of Kovalam Beach*. Unpublished MPhil Thesis, University of Kerala, Thiruvananthapuram.

Waters, M. 1995. *Globalization*. London: Routledge.

Wolf, Y. 1993. The world of the Kuta cowboy. A growing subculture of sex, drugs and alcohol is evident among male youth in the tourist areas of Bali and Lombok as they seek an alternative to poverty. *Inside Indonesia*, (June), 15–17.

Chapter 4
Neoliberal Development and Displacement: Women's Experiences of Flight and Settlement in South Asia

Ruchira Ganguly-Scrase

Introduction

This chapter explores the experiences of poverty-induced internally displaced and undocumented cross-border migrant women from Bangladesh living in the Indian state of West Bengal.[1] Such a comparative analysis is valuable since the underlying reason for their displacement is the same, namely, neoliberal development under a globalising economy. However, rather than positing a simple causal relationship between neoliberal globalisation and dispossession, I suggest that globalisation has accentuated the conditions that lead to displacement. While globalisation conjures up a vision of a borderless world, as a result of free flow of goods, increasingly the Indian nation state has hardened its attitudes towards the displaced, often becoming mired in debates around 'foreign infiltration'. The failure of the nation state to manage ethno-cultural justice for minorities and poverty alleviation are but two sides of the same coin. Moving away from statist models of modernisation and development and embracing neoliberal agendas of privatisation and free markets has resulted in devastating consequences for the poor. Based on fieldwork among displaced women, this chapter outlines how women and their families cope as forced migrants and how women themselves assess their situation. In doing so the gendered complexities of dislocation are highlighted; additionally critical attention is drawn to the weakness of the concept of refugee in contemporary post-colonial contexts. By interrogating the complex relationship between gender and displacement, marked by unequal economic and cultural domination, I question the value of distinguishing between 'economic migrant' and 'political refugee' for understanding women's settlement experiences. It has been argued that for many postcolonial states, a place in the global economy means that the contradictions between capital and labour must be constantly kept in suspension, reducing the nation's capacity for independent development (see Burkett and Hart-Landsberg 2003; Kaplinsky, 2005; Rudra 2002). In both India and Bangladesh, the pursuit

1 This research was supported by URC grant, 'Gendered Exclusion' University of Wollongong.

of adjustment policies over the past two decades has negatively impacted on the poor. Therefore, I will firstly consider the problematic relationships between neoliberal development, globalisation, and its particular manifestation with respect to displaced people in India. This will be followed by an overview of how displacement affects women. In the final section, I will present an ethnographic case study of displaced women in West Bengal.

Neoliberal Globalisation, Adjustment Policies, and Dislocation

Forced migrations are a fundamental part of globalisation and, thus, cannot be studied in isolation. According to McNevin (2007) geographies of neoliberal globalisation are implicated in irregular migration flows. Refugee situations are not a string of disconnected humanitarian emergencies and are connected to a wider social, political, and economic context. By considering the broader structural causes of forced migration, one can generate explanations for its rise (Castles 2003a: 17–30). Globalisation is characterised by elements which are both neoliberal and neoconservative. They share more similarities than differences (Robison 2005; Steger 2005: 17). Both neoconservatives and neoliberals emphasise the significance of free trade and markets. Neoconservatives, however, also combine this attitude with a belief in the regulatory actions of governments and in the protection of their citizens, in terms of both security and traditional values. Yet, their notions of security do not imply human security as understood by Human Rights discourses (cf. Annan 2005; McAdam 2008) Instead, as both Harvey (2005) and Peck (2004) argue, neoliberalism is not merely the simple application of free market philosophy, but is based also on social conservatism and increasing preoccupation with social control in areas of law and order and border protection (Peck 2004). As far as security is concerned, neoconservatives may even differ from neoliberals in their approach to the role of the state. As Akram-Lodhi (2005: 164–65) notes rather than rolling back the state, neoconservatives are willing to countenance budgetary deficits in areas, such as security. Instead of being the borderless world characteristic of free trade and of transnational elites, globalisation is in fact about borders which are both permeable and exclusionary (Benhabib 2005; Rudolph 2005).

Evidently, forced and economic migrations are closely related and often interchangeable expressions of global inequality and societal crisis (Castles 2003a, 2003b; Cohen and Deng 1998; Weiner 1996). It is through the deconstruction of various bureaucratic categories that both the diversity and similarity of people's experiences can be exposed. According to Sivanandan (2001: 87), the distinction between the political refugee and the economic migrant is a false one and is vulnerable to differing interpretations depending on the interests of who such categories serve. It is the interests of the powerful that have resulted in the blurring of these categories. Neoliberalism with its focus on structural adjustment programmes has resulted in reduced social spending, leading to the impoverishment and eradication of social, welfare, and educational provisions to people in developing nations. Resistance to poverty cannot be separated from political resistance and persecution (Lahiri-Dutt and Samanta 2004; Lindio-McGovern 2007), thus, turning the political refugee into an economic migrant. In addition, millions of people become displaced each year as a result of development programmes (de Wet 2006; Robinson 2003).[2] It is normally difficult to tell the difference between environmental, economic, and political factors and, therefore, the category – environmental refugee – can obscure the very complex reasons underlying environmental disasters. Often, the underlying causes of such forced migration might be found in the chosen path of development followed by the state (Mishra 2004).

Neoliberal globalisation has adversely impacted on women by exacerbating poverty in developing countries (Elson 2002; Razavi 2002). In countries like the Philippines, female labour export, especially of domestic workers, has been a government strategy to tackle the negative impacts of neoliberal policies. Yet, there are disproportionately high numbers of undocumented Filippina domestic workers abroad without appropriate security; similarly, growing poverty has led to trafficking (Lindio-McGovern 2007). With respect to India, the New Economic Policy introduced in 1991, after pressures from the International Monetary Fund (IMF) has resulted in growing pauperisation, which leaves very few resources for girls' schooling. A number of scholars have shown the devastating impact of the New Economic Policy on women in India stemming from commercialisation of agriculture, privatisation of public sector enterprises, reduction of subsidies, and poverty-eradication programmes (Batliwala and Dhanraj 2004; CWDS 2000; Upadhyay 2000). Correspondingly, according to Naruzzaman (2004: see especially pp. 39, 44, 48, and 51), Bangladesh's ambitious, aggressive, open market, and privatisation policies have not only resulted in increasing inequality between the rich and the poor, but also that women have suffered the most as the resulting privatisation of industries enables employers to fire women very easily. Kabeer and Mahmud (2004) present a more complex and nuanced account of the gendered nature of Bangladesh's production for the global market and its impact on poverty alleviation. Nevertheless, they also point to the ruthless exploitation of women from the poorer strata. Subsequent violence against women is also on the rise (Khan 2005).

2 Although this chapter is not directly concerned with development-induced displacements, it is worth noting the enormity of the problem in India. It is estimated that in India during the last 50 years, 25 million people have been displaced by development projects (Robinson 2003: 3). Moreover, as Basu (in this volume) shows, it can be catastrophic when a state targets a particular section of the population to beat a disproportionate share of the costs of development, and through either neglect or outright malice, denies them a proper share of its benefits. Further, the largest victims of internal displacement and development-induced displacement are women and children (Robinson 2003: 17). It is often the ethnic 'otherness' of victims of development that takes the pressure off nation builders (Robinson 2003: 11).

In addition to the increased the impoverishment of women which makes it difficult for them to find work, there is ample evidence to show that globally neoliberal prescriptions have been often accompanied by aggressive border protection policies (Kipnis 2004: 259–64; Overbeek 2002; Sparkle 2006). In postcolonial states, the resultant impact has affected women even more adversely than its effect on men. The Indian subcontinent exemplifies the quintessential artificial boundary construction across diverse communities in process of postcolonial nation-making. Homogenisation underpinning nationalist ideologies also led to the creation of marginalised and displaced 'minority' groups who did not belong in the nation state. Contestations over artificially constructed boundaries have continued since simple movement of populations in border regions now becomes an illegal act. However, the pursuit of market liberalisation in India does not seem to be leading to greater tolerance of ethnic and religious differences. On the contrary, border controls are being vigorously reinforced to keep alien *others* out.

Nation-making, Border Constructions, and Displacement

India is emblematic of the mass displacement prevalent in the developing world. Although India is a common place of resettlement for the displaced of nearby countries (Nair 1997), ironically, no coherent policies have been developed. Not a single South Asian state is signatory to the United Nations (UN) convention on refugees (Ahmed et al. 2004: 4). Complexity arises because of the legacy of partition. In the aftermath of Partition, while refugee rehabilitation was more thoroughgoing in Punjab, in light of population exchanges, in West Bengal, the movement of populations was episodic and financial resources allocated for rehabilitation far less (Kudaisya and Tan 2000). More significantly, tension and contestation over artificially drawn boundaries arise, resulting in blurring the boundaries of the experiences of those displaced internally and internationally (see also Ahmed this volume). The very notion of displaced persons has been called into question by an examination of the diverse forms of displacement in northeast India. This is outlined by Samaddar's (1999) reconceptualisation of the distinction between 'migrant' and 'refugee', which posits the nexus between the two terms and the shared common factor of victimisation. Samaddar (2003: 11) argues that giving different definitions to populations is a meaningless exercise as 'in reality … differences are fruitless'.

The arbitrariness of post-partition Bengal has had profound consequences for the women in this study. This process has been highlighted by Feldman, who notes that almost overnight there was a shift from shared 'understanding and identification with a national project' to a distinction that for many 'verged on the nonsensical' (2003: 113). To this one can add Van Schendel's (2003) assertion that working people's livelihoods were drastically affected after Partition. His findings are of critical significance for understanding the sentiments of many of my informants in this study. While the Hindu and Muslim elites either exchanged properties or were compensated for their losses following partition, the same did not apply to the property-less or the poor. This affected those living in the border regions the most. For many of them, seasonal migration was a routine practice in times of hardship. However, this opportunity was foreclosed at the stroke of midnight in August 1947 as the same movement of people across the border was rendered illegal. Yet, internal displacement continues to be an acceptable phenomenon in both nations and has been exacerbated by the current neoliberal adjustment policies.

Borders between nation states in South Asia are largely artificial constructs arbitrarily drawn through ethnic, religious, cultural, and economic communities. Such borders can be found, for example, between Kashmir and Pakistan, China, and India. The artificiality is exemplified in the state of Tripura where a border divides a football field (Acharya et al. 2003: 6). The fluidity of borders, which often isolate communities, is highlighted, for example, where a river that often changes course is a national border. Despite espousing the benefits of border trading and some recent attempts at regional economic cooperation, free movement of people is yet to be achieved. I shall return later to the tension and contestation that arise over artificially constructed boundaries since they are relevant to the discussion of the experiences of women in my research.

Rigid Categorisation and Its Consequences for Displaced Women

Historically, the conceptualisation of the displaced person can be traced to the inception of the UN Convention of 1951 within the context of post-war reconstruction. Here, the definition of a refugee was characterised as a person who, 'owing to a well founded fear of persecution for reasons of race, religion, nationality, membership of a particular group or political opinion, is outside the country of his nationality and is unable, or owing to such fear, is unwillingly to avail himself of the protection of that country' (Article 1, United Nations Convention Relating to the Status of the Refugee, 1951, cited in Loescher 1999: 234). This definition is both antiquated and insufficient in the twenty-first century (Suhrke and Newland 2001: 284). With its focus on individual persecution and sovereignty, this notion is Eurocentric and reductionist (Loescher 2001; Malkki 1995). 'Economic migrants' are not recognised as *bona fide* refugees because they are assumed not to suffer from 'persecution', but are said to have an element of choice in their movements (Lundquist and Massey 2005; Stevens 2003). The global community eschews responsibility for the internally displaced, under the guise of observing state sovereignty. According to Phuong (2005:73) only a minority of the 20-25 million internally displaced persons across the globe are assisted and protected by the UNHCR. Forced displacement has only recently been perceived as a human rights problem (Stravropoulou 1998: 519: Bayefsky and Fitzpatrick 2000; McAdam 2008). Conventional demarcation between displacement due to coercive measures (war, direct persecution, famine) and displacement due to economic reasons (poverty) glosses over the fact that both categories of displaced persons often suffer under the same abject conditions after relocation (Hein 1993: 47). These aforementioned mass persecutions and displacements exemplify the limitations of conventional refugee discourses that are predicated on individual persecution. Arguably, the refugee paradigm also excludes the experiences of women in forced internal and international migrations.

Women: The Invisible Refugees

The stereotype of a 'refugee' conforms to the Cold War image which is predominantly male. According to Haines (2003), the language of the UN Convention on Refugees is gender-blind. Women have often been relegated to the periphery in mainstream refugee debates. Arguably, women have been excluded from both internal and external displacement debates due to the very definition of what it means to be a refugee. Some scholars argue that women are systematically excluded from the refugee debate due to the specificity of female experiences, despite the overrepresentation of women in refugee statistics (Boyd 1999; Cohen 2000; Macklin 1995). Boyd's (1999) incisive analysis of the gendered nature the refugee experience is worth re-counting in detail. She contends that the United Nations High Commissioner for Refugees (UNHCR) definition of the 'refugee' is simultaneously individualistic and presumptuous in its intimation that violations must be specifically committed by the state (Boyd 1999: 8). Boyd suggests that this definition privileges the public side of public/private divide by focusing on the actions of the state. It 'fails to acknowledge forms of persecution that occur in private settings' (Boyd 1999: 8). As women are more likely to be persecuted in the private sphere, they are less likely to be officially seen as refugees.

It has been argued that refugee law is intrinsically gendered and subsequently needs to be altered (Macklin 1995: 218). Further, Boyd posits that female persecution, eventuating in displacement, can be conceptualised in two ways. First, a woman can be persecuted as a woman, not because she is a woman. This means that the form of persecution is gendered, such as in the case of rape. Correspondingly, a woman can be persecuted because of her gender or because she has broken social mores pertaining to her gender (1999: 9). This distinction is made clearly by Haines (2003) who describes the former form of persecution as gender-specific and the latter as gender-related. Citing Crawley, Haines (2003: 336) also describes persecution by sexual violence as a weapon of war when the '… violation of women's bodies acts as a symbol of the violation of the country'. Violence against women is regarded as a 'natural' circumstance of war and conflict. Though rape is recognised as a war crime, it is not universally accepted as a ground for refugee status, as noted above. Pittaway and Bartolomei (2002: 26) illustrate the illogicality of this with the example of a woman who was not granted refugee status because she had been raped, although her husband was accorded this status because he had been forced to watch her being raped.

Gender inequity and stratification can be reproduced in places of relocation where women generally possess less education and fewer skills than their male counterparts, and, hence, lack bargaining power in the community (Boyd 1999: 12). Another important aspect of this analysis is the way in which women's bodies often become sites of contestation in relocated communities. The degree to which women are controlled is often a symbol of reconstructed patriarchal authority in many displaced communities (Ganguly-Scrase and Julian 1997: 435).

Gender Relations, South Asian Cultures, and (Re)settlement

In general, gender relations in South Asian cultures are such that experiences of women are rarely given the same significance as that of men. In displacement, the loss of property and work, physical injury, separation from family, and issues of protection and security are all more serious for women (Basu Roy 2002). Bose (2000) and others argue that during the Partition of India, women became the symbols of the nation and had to be protected because national pride was at stake. At this time, abducted women were repatriated back to families and women were killed by their families if it seemed likely that they would fall into enemy hands (Butalia 1998). Women not only were symbolic of the nascent nation, but were also the carriers of culture and tradition and were, therefore, precious. Several decades on these values persist.

Asha Hans (2003: 379), the well-known South Asia Refugee Rights activist and researcher on humanitarian protection, states that 'gender consideration was never an important component of India's refugee policy'. Despite the fact that displacement and asylum is a gendered experience, women are seen as 'objects not subjects of humanitarian planning' (Hans 2003: 380). Subsequently, Banerjee 2002: 9) cites Das (Samir Das, "Ethnic Assertion and Women's Question in Northeastern-India," A.K. Jana, ed., Indian Politics at the Crossroads (New Delhi: 1998) p.177.) in highlighting that the 'South Asian attitude to women has been guided by "mystified notions of chastity"' which leads to the notion that women in South Asia belong to their own communities. To this she adds when women are displaced in large numbers, the focus shifts 'from the individual woman to their communities' (Banerjee 2002: 9). The guidelines for the protection of women are often left to individual governments to put into practice and where governments are gender-blind, these guidelines are not put into practice (Banerjee 2002: 9).

Women and Forced Migration: Similarities and Differences between Refugee Women and Poverty Induced Displaced Women

Globally women and children constitute 80 per cent of refugees and internationally displaced people (Rodriguez 2003: 6). Women experience gendered forms of violence, such as rape, the fear of rape, body searches, enforced pregnancy, slavery, sexual trafficking, enforced sterilization, and infection with sexually transmitted diseases and AIDS, as well social stigmatization once they have been sexually assaulted (Rajasingham-Senanayake 2004: 149; Rodriguez 2003: 1). Sexual violence, including the infection of women with HIV/AIDS leads to local discrimination against these women, which leads to discriminatory practices against them in resettlement. Among the internally displaced, similar patterns can be found. For example, Banerjee (2005) provides one of the most extensive accounts of gendered internal displacement throughout South Asia. These include rapes by security forces in Sri Lanka as an instrument of displacement, systematic use of rape by the Burmese military as a weapon against ethnic minorities, police collusion in genocidal acts against Muslim women by militant Hindu mobs in Gujarat, Bangladeshi state-sponsored terror against Jumma communities in the Chittagong Hill Tracts that has led to abductions, kidnappings, and forced marriages of young women.

Children's survival often depends on a woman's ability to adapt to impoverishment. A woman fleeing from hardship, violence, and war faces the threat of rape by the border guards; she is also encumbered with the task of childcare, cleaning, cooking, and collecting fuel and water. The collection of fuel and water often leads to further violations of her body and soul (Samaddar 1999: 41). During fieldwork, I found that women are expected, even under the conditions of forced displacement, to take care of the family and to uphold cultural traditions. This expectation holds even when women are abandoned by their husbands and thus denied traditional protection, left without a home, possessions or work, and without any family or community support. Such patterns are common not only in South Asia, but also hold true for patriarchal societies generally. Samaddar's (1999: 40) findings on South Asia shows that women are the most abused refugees and the most unwanted migrants. The sexual victimization they face is the most gender-specific human rights violation of forced migration. Women experience greater poverty, health risks, have more mental health problems, receive less information, fewer work opportunities, and fewer opportunities for education and training than men when relocated (Banerjee 2005; Kaapanda and Fenn 2000; Waas et al. 2003: 330). In addition, women frequently face social stigma if they are living alone, or accusations of promiscuity (Qadeem 2003).

Close similarities between female refugees and poverty-induced displaced women can be found throughout South Asia. Later in the chapter, through women's narratives of displacement, I shall highlight the ways in which the boundaries of the economic migrant and the political refugee became blurred. As noted earlier, both groups of women in my study are located in a region where an artificial international boundary was imposed overnight. Subsequently, while displacement within the national boundary has become acceptable, the movement of people across the border is rendered illegal. Yet, both groups continue to suffer from economic deprivation and policies that systematically exclude women from outside assistance.

Similarities between female refugees and poverty-induced displaced women are also evident when examining the way in which aid organizations liaise with male figureheads, who are often unaware of specific female requirements (Okin 2003: 280; Ganguly, 2005). In both forms of displacement, which usually entails scarcity, women are generally given fewer resources, such as food (Okin 2003: 281). In displacement caused by development projects, gender biases continue to persist, particularly in relation to re-settlement programmes (Bisht 2009; Mehta, 2009). Cash compensation is given to men, and women are doubly disadvantaged in that they lack economic capital in addition to land and skills they had developed particular to that locality (Arora 1999: 345).

Research Method: The Importance of Ethnography

My research is based on an ethnographic approach. A number of scholars have highlighted the significance of this method for understanding forced migrations (Colson 2003; Malkki 1995; Sorensen 2003; Eastmond 2007). Long-term intensive fieldwork, the hallmark of this method and participation in the everyday lives of the displaced, enabled me to interpret their lived experiences that are embedded in a web of complex social relations and historical processes. The findings for this chapter are derived from fieldwork among forced women migrants in 2004–2006 in Kolkata in the Indian state of West Bengal and Siliguri in the northern region of the state, including in-depth interviews with 13 key informants. Bangladeshi respondents are both legal and irregular immigrants. The other group comprises women who had been internally displaced within the state of West Bengal due to poverty. The women's ages range from 16 to 75 years and include single, widowed, married women, and deserted wives. My observations and analysis are informed by intensive and intermittent fieldwork carried out in this region for nearly two decades. The first period of intensive fieldwork was carried out for 18 months in Nadia, a district bordering Bangladesh, the second largest settlement of refugees.

A number of my current key informants are those with whom I have maintained a close and continuing relationship since the early 1990s. Fieldwork conducted in this region over a prolonged period among various communities and classes enabled me to form close links with a number of displaced Bangladeshi women. Such contacts allowed me entry into networks of dislocated communities. This access was critical since emerging hostilities towards alien 'Others' stemming from changes in the political climate in recent years have made refugees and newly arrived undocumented migrants fearful and reluctant to talk to outsiders. Gaining their trust was vital in uncovering the detailed accounts thus presented.

It is important to note that those who had migrated from Bangladesh did so in different waves under different circumstances. The women's experiences varied according to their ages and the then socio-political problems, commencing in the 1950s in the immediate post-Partition years when they were teenagers to those in the 1970s during the Bangladesh War of Liberation when they turned young adults.

A number of women belonging to the latter age group went back to Bangladesh, only to return in the 1980s and the 1990s. Most recent arrivals reported persecution and harassment by the dominant majority, as well growing impoverishment.

The experiences of women in this study are part of a larger, ongoing, cross-border fluid movement of populations into northeast India. During colonial rule, the seeds of dissension along communal lines were sown way back in 1901. The province of Bengal was partitioned into East and West, with Muslims predominating the former and Hindus the latter. At partition in 1947, with the two nation states being carved out, namely, India and Pakistan, a large majority of Hindu Bengalis arrived as refugees in West Bengal. Yet, a significant proportion continued to live in East Pakistan (now Bangladesh) as their identification was along local cultures rather than nationalist identities. Bengali cultural nationalism, drawing on secular principles and regional identity, intensified in East Pakistan against Punjabi-dominated exclusionary politics of Pakistan, culminating in the war of Independence and the formation of Bangladesh as an independent country in 1971. More recently in Bangladesh, growing instability, poverty, and militarisation have given rise to Islamic fundamentalism (Jahid 1996; Lintner 2002) progressively pushing out Bangladeshi Hindus (Amnesty International Report 2001) as well as the Muslim poor. In West Bengal, in the past half century while cultural politics and public culture has been shaped largely by secular and democratic principles, often spearheaded by the post-partition 'East Bengali' refugees, fractures are also appearing as a consequence of assaults on two fronts: Islamic fundamentalism of Bangladesh and growing Hindu nationalism of greater India. Until its demise early in 2011, West Bengal's ruling Left-front government and especially the Communist Party of India (Marxist) [CPI(M)] has consistently maintained an anti-communal stance and enjoyed popular support due to its pro-poor policies for the past three decades. At the same, however, the pragmatic approach of expanding its support base by incorporating immigrant Bangladeshis into the electoral list and thereby indirectly engineering citizenship rights is a source of intense hostility. Paradoxically, however, by adopting neoliberal agendas, it is unable to reconcile its traditional pro-poor rhetoric with the reality of mass displacement emerging from urban redevelopments which favour the new middle classes. Similar contradictory phenomenon can be observed in relation to the estimated 16 million Bangladeshis who have moved across the border over a period of several decades and are now being considered a security threat (Kumar 2010; Singh 2002). The reconfigured discourse around security is symptomatic of a broader global trend. There has been a shift from granting sanctuary or supporting the rights of the dispossessed to that of demonising them as a threat to law and order. While discourses of 'foreign infiltration', initiated primarily by Hindu nationalists have dominated public debate in West Bengal in recent years, in the surrounding states, the resentment is far more violent.[3] The CPI(M) has also played a dubious double role in recent

3 The Bharatiya Janata Party (BJP), the pro-Hindu nationalist party and the Hindu Right in general speak in terms of Muslim Bangladeshi immigrants as 'infiltrators' who

years. For example, in 2008, displacement caused by development plans resulted in violent acts against one of its traditional constituencies of Muslim peasantry on the urban fringes. In this situation, the state government tried to partly absolve its abysmal failure to meet the needs of its traditional support bases and also to appease Islamic conservatives by extraditing from West Bengal the outspoken Bangladeshi feminist author Taslima Nasreen. How some of these issues have the shaped the experiences of women are presented now.

Women's Accounts of Displacement

Nation and Justice

For most, the concept of nation is largely a farcical abstraction. In routine conversations, such as discussing hardships or their longing for home, the term *desh* was used to describe their place of attachment. In everyday language, *desh* literally means a country, motherland, or one's native village. However, it is also synonymous with the term 'nation' or 'nation state'. When asked what action their families would take if survival in the village became unbearable, a typical response from the internally displaced women was:

> When it would become very difficult they would leave the *desh* and go to *bidesh* [elsewhere; literally 'a foreign place'] and would do some kind of work. And those who would remain behind they would go out and work and try to earn a living for their survival. (Married woman, periodically separated from husband, aged 30. In-migrated from Midnapore District)

'Bangladesh' is, thus, an apt name for the land of the Bengalis. Samaddar (1999: 79) makes a similar point concerning 'home'. Reporting on immigrants' accounts, he shows that desh is where they come from: Faridpur or Khulna or their village of origin. He argues that to them the liberal notion of citizenship has no particular meaning. Rather, it is constitutive of functional requisites, such as voting rights, access to public distribution systems, and so on.

It is only ideological rhetoric that invokes the difference of the minority other. This is generally illustrated in problems faced by Hindus staying behind in Bangladesh. The case studies have revealed absurd situations of Hindu families that had lived for generations in Bangladesh among Muslims now experiencing

must be driven out from India, while Hindu cross-border migrants are regarded as bona-fide refugees (Singh 2009). These claims are more pronounced outside of the state of West Bengal. See Gillan (2002) and Ramachandran (2002). In the northeastern states of India, the nature of hostilities directed against Bengali immigrants are an outcome of a much more complex history whereby the composition of local ethnic and tribal populations have radically transformed by in-migrating Bengalis, both Muslim and Hindu (Bhaumik 2000) abusive language from their neighbours and being told to 'go to India'. Thus, in the construction of national identity, Hindus from Bangladesh are placed in the precarious position of being the 'other' (Feldman 2003: 112) in both India and in Bangladesh. They are an Indian in Bangladesh, and a Bangladeshi in India. As one woman commented:

> People who have come to India in recent times are called Bangladeshis. The real life difficulties remind us that we are Bangladeshis ... all this reminds us that we are different and we are Bangladeshis. But we never consider ourselves Bangladeshis. *We don't have any attachment with Bangladesh.* (Emphasis added)

While I do not wish to make too much out of this statement about national sentiments and attachment, since many of the undocumented migrants are desperately struggling to stay out of trouble and giving a routine response to an inquisitive researcher to gain sympathy, it does illustrate their precarious status as minorities and being poor. Here, I am also reminded of the paternalism of the dominant classes towards indigent cross-border migrants and refugees. In the early 1990s, while carrying out fieldwork in Nadia district (where large numbers of Bangladeshis have settled), many of the genteel middle classes, though sympathetic towards the displaced, would nevertheless point to their otherness. The newcomers' cultures eroding the town's identity and the sense of values would be frequently expressed by evoking nostalgia for a bygone era:

> The small local population of pre-partition days virtually knew each other. There was a sense of values [*mulyabodh*]. The families were connected. The new population, stigmatised by their 'refugee' label today outnumber the original inhabitants. They have faced enormous social stress as a result of partition, they battle to survive. The uprootedness has distorted their way of life and outlook. It has fostered *other* ways of thinking. [Emphasis added)

It goes without saying that minority persecution is most acute in the cases of refugee women I spoke with. This can be understood in the instances of threats to violation of their purity as a tool of minority oppression. However, this is a persistent phenomenon concerning gender relations in South Asia and is not confined to any one community (Sunder Rajan 2000). Nevertheless, it is worth noting women's own narration of the complexities of their experiences that forced them and their families to migrate to India. Some categorically stated their minority status: fear of religious persecution compelled them to leave their homeland and find shelter in another country. Yet, despite their reference to communal tensions being a cause for their dislocation, they did not personally experience physical attacks. Only those with small business and property faced direct threats. What they did acknowledge was that there had been a constant fear in their minds given that they had young daughters at home. They had heard of incidents whereby Muslims had abducted Hindu girls or even raped them. It is interesting to note that while they recounted such details of Muslim atrocities happening somewhere in their country, in case of their own safety, they were grateful to the Muslim neighbours who had protected them.

Such views are evident from some of the following comments made by few women:

> Hindu women and grown-up girls had to live with a sense of insecurity. We normally avoided venturing alone in places outside our house. Women and girls from our community were constantly under threat. There were incidents of sexual harassment by Muslim youths..... But we were not physically attacked. There was fear of an attack always.... (Deserted woman, aged 40. Migrated from Natore, Rajsahi district, Bangladesh)

This is remarkably similar to the accounts of earlier post-Partition refugees:

> About Bangladesh, I can only tell you about riots.... When it started, I must have been about 9–10 years old. Some of it is a blur now.... Oh! It was because of the riots that we had to leave.... We were chased by the Muslims, we ran here and there... We were hearing that this one's daughter was taken and cut up in front of the mother. But as far as we were concerned, we were alive because of the Muslims' help. Yes, the Muslims saved us. There were good and bad among them.... (Married woman, aged 75. Migrated from Narangar, Dhaka, Bangladesh)

Few recently arrived refugees (1990s–2001) equally attributed poverty as a reason for migration. Such views were typified by comments such as:

> ...Then after the trouble, we came here because of persecution by the *miyas* [Muslims]. It is because of poverty that we had to leave...It is because of persecution and poverty that we had to leave, I mean if we could not gather grains, then how can we bring up our children? (Married woman, aged 40. From Khulna district, Bangladesh).

The accounts of economic dislocation and the reasons for migration did not differ in any way from the internally displaced. In several 'refugee' narratives, I could not identify anything that separated them from the internally displaced women of West Bengal. The following contrasting examples illustrate this point:

> I made it because my husband was no longer there. My brother-in-law had just got married. Then he started creating trouble, he would not feed us. That's why we left. When his father died, then my elder son, I had a sister-in-law here, so we came to our sister-in-law's house with my mother-in-law. My sister-in-law put us up. (Married undocumented migrant woman, aged 38)

and:

> Yes, I came here. Then my husband objected to it, asking, 'Why did you come here? You must stay in the village'. But how was I to stay there? How would we survive? I said to him, 'You don't give me any money, what am I to eat?' He told me that as the others eat once a day only, I should do the same. But you tell me how could I do that, especially with the children? I have been living here now for nine years. (Internally displaced woman, head of the household, aged 37)

However, it would be too simplistic to suggest that greater feeling of security (or being grateful to have a home and livelihood) implied identification with the culture of the majority in West Bengal. A 16-year-old, who worked as a seamstress was quick to point to the difference:

> I didn't know what I was going to expect. I came here with my parents. I didn't think anything. I don't know. People here are all right. They are different compared to [those in] Bangladesh. They are different they way they talk … their character is different … they're not very nice sometimes. I feel homesick, very upset sometimes. I try very hard here, but we don't have any money, so can't go back.

This reinforces my earlier point of the ambiguity of otherness.

With the exception of one respondent, who mentioned moving across the border in a group of 50–60 people, most of the refugees had migrated with their families. From the discussions with the Bangladeshi immigrants it can be inferred that reasons for migration were not merely political and religious persecution as minorities, but economic as well. Women did not hold any individual or community responsible for their dislocation. They simply attributed it to their fate or God's will or to a decision of some family member of theirs. While some (mostly Bangladeshis) referred to sudden flight, others (particularly IDPs) said that it had been a gradual process that had been well thought-out. The latter unambiguously explains that poverty and poor economic conditions compelled them to migrate from their villages seeking better employment opportunities in the city: 'In the village, there was no rain, there was drought and poverty. So we came to Kolkata' (Deserted woman, aged 40. Migrated from village Lakkhikantapur, West Bengal).

However, most emphasised the absence of tension between communities along religious lines in their native villages. A typical example of the routine description of relations among villagers is presented below:

> And Parbati, all these communities in the village, was there any enmity among them?
> **A:** No, there is unity in our desh [home]. Muslims and Bengalis, we were all united.
> **Q:** Ok, so Muslims and Hindus, and amongst the Hindus, all those *Napits* [barbars], *Dhopas* [washermen] all those different …?
> **A:** All, all, we were the same.

It is impossible to say whether such emphatic claims about the unity of all communities, of class alliance, especially of the poor, across ethnic and religious divide narrated by women from rural West Bengal are a genuine reflection of their belief in the CPI(M) ideology or simply a successful reproduction of CPI(M) rhetoric. On the one hand, suspicions can be raised from the separate categorisation of 'Muslim' and 'Bengali'. On the other, one may wonder why is there the need to give such a stock answers, even in routine conversations.

Growing class inequalities and mobilisation by the CPI(M) cadres has facilitated a shared degree of identification between West Bengali and Bangladeshi displaced women. Both groups provide essential urban services to the middle classes, but are frequently labelled as 'undesirables'. Bangladeshi women were further demonised as 'illegals' and 'infiltrators.' However, despite the implementation of neoliberal policies that aim to privatise public space and growth of shopping malls, Indian elites are still compelled to share public spaces with those Kaviraj (2000: 241) terms as their 'habitual inferiors'.

Kin Support and Survival

Women were under pressure to support their families often assuming the role of breadwinners. Yet, none except one woman noted have received assistance from any official sources. Bangladeshi women could not to have survived without traditional kinship obligation being exercised. Apart from kin, the only outside assistance in the form of non-financial help, was from Party cadres, which appears to be more prevalent in north Bengal than in Kolkata. While internally displaced women reported being constantly harassed by local hoodlums in their neighbourhoods (though they were grateful to have police protection), refugee women, particularly those in Siliguri, feared the police. The latter faced the constant fear of eviction for which they had to bribe the police. Similarly, Bangladeshi women's positive appraisal of the freedom of female mobility, their public visibility, and their subsequent absence in Bangladesh contrasted with internally displaced women's more reserved and circumspect observations. These differing views were in part shaped by Bangladeshi women's sense of security living among a network of kinfolk.

Development has had a limited impact on addressing the needs of women, generally failing to improve their quality of life and their status. Today, only 30 per cent of the world's poorest people are male. In rural areas, the West Bengal government policies have enabled some of the poor, including the brothers of some informants, to purchase their own land. Yet, the findings of this study reveal that women remained dependent on their male kin, both for land and the cultivation of that land. A number of women in my study faced starvation and, subsequently, were forced to move to the city. This phenomenon is further evidence of research carried out by Hill (2003) which shows that despite the espousal of pro-poor policies, the state has failed to ensure the well-being of the very poor in West Bengal.

A life of hard labour is something that many women share. Currently, the rights of women to leisure as guaranteed in the UN Charter of Human Rights (Article

24), are being severely undermined by a failure to prioritise labour rights[4] and a patriarchal understanding of labour (Niewenhuys 1994: 23; Vervoorn 2006: 71–73). Therefore, although women's informal sector work takes up the major proportion of their time, there are currently no laws to protect women against the unfair exploitation of this labour. An example was the case Parbati (her comments have been recorded earlier in this chapter), whose husband though he relied on her unpaid work, refused her adequate remuneration even to sustain her. Even when these women did shift to paid employment, the financial reward was painfully inadequate since the only work they could secure was in the unregulated 'informal' economy. It is important to note that while undocumented migrants mentioned the difficulty of obtaining ration cards, which they regarded as the key to social justice, those possessing these cards felt that a better income would improve their lives. While the middle classes under economic liberalisation have experienced exponential growth in their incomes, the respondents who worked for them still only earned between Rs 300–1,200 per month. The nature of globalised development detracts attention away from protecting women's rights, as women's increased work hours are a key mitigator of economic liberalisation, structural adjustment, and other austerity measures that have been forced upon them by the self-appointed champions of human rights (Rai 2002: 144; Vervoorn 2006: 71–73).

In the final analysis, women's lives demonstrated me the ambiguous and complex way in which migration can be much more painful for women, and at the same time-empowering. They all mentioned the normative practices of the past: women did not have to participate in work outside the home. In the case of some undocumented migrants, this was made possible by a relative affluence of the household. It is still a practice that is looked down upon, but the failure of their husbands to provide for them forced them to do so. All the women, due to circumstances beyond their control, were forced to migrate and go against familial norms. However, all experienced a degree of empowerment, and, as a consequence, becoming the 'breadwinner' and sometimes the head of the household, and as a result of separation from elder kin. Some enjoyed the freedom of mobility in their place of settlement. One woman described:

> In Bangladesh, we never moved around freely; life was confined within the walls of the house. I could never think of doing a job outside my house in Bangladesh. Besides there was ever present threat of physical [sexual] assault at the hands of local Muslim youth. In West Bengal, there are no such threats. The atmosphere here is easier for women. Women can be seen everywhere in the markets, at the workplace and in the fields which was not the case over there. (Deserted woman, aged 33. Migrated from Rajsahi district, Bangladesh)

Despite experiencing greater hardships than men, my research shows that displaced women are active in making use of whatever limited resources that

4 Labour rights are rarely couched in Human Rights terms. See Fukuda-Parr (2006)

are available; finding food and shelter for their dependents. Paradoxically some aspects of displacement have facilitated women's capacity to assert a degree of independence that may not have been possible had forced migration not taken place. This corresponds with the findings of Manchanda (2004) on IDP and forced migrant women in South Asia, which shows that long-term displacement also provides women the opportunity for greater personal and group autonomy.

Conclusion

By comparing a number of areas in women's everyday lives, such as nature of home village and childhood, reasons for flight, settlement processes, subsistence and survival, safety and security in areas of settlement, personal and national identities, attitudes towards locals, and the desire to return, the observations made in this chapter demonstrate that there are close similarities between female refugees and poverty-induced displaced women. My informants continue to suffer from economic deprivation and are systematically excluded from outside assistance.

Although women are commonly exposed to many vulnerabilities, they face regular threats of violence and cope with reinforced patriarchal authority, this research thus far also reveals that there are possibilities of their empowerment. Most women valued their freedom of mobility and enhanced sense of self-confidence that the new environment has presented them. However, for women who have arrived as undocumented migrants, their awareness of being non-citizens is acute. They openly speak about their sense of being deprived the rights of citizenship. For example, they find it difficult to obtain ration cards, enrolling their names on electoral lists, and even in admitting their children to schools (where they are asked to produce birth certificates, using the rhetoric that birth registration to be legally mandatory for every child). They feel disadvantaged for not being Indian citizens, particularly due to the continual harassment by the local police. They compare their situation with earlier migrants (some of who are fellow villagers, caste fellows, or relatives), who have now established themselves in economic, social, and political arenas.

Research in peripheral states elsewhere suggests that the imaginary community of nation states has become tattered at its edges, but the pursuit of market liberalisation does not seem to be leading to greater tolerance of ethnic and religious differences (Hann 1997). On the contrary, in South Asia, ethnicised models of religion are being utilised to reinforce strict border controls, while simultaneously promoting a vision of regional trade liberalisation. Ultimately, neoliberal development has failed to address the needs of the internally displaced and cross-border migrants seeking refuge.

References

Acharya, J.M., Gurung, M., and Samaddar, R. 2003. *Chronicles of a no-where people on the Indo-Bangladesh border*, SAFHR Paper No. 14, South Asia Forum for Human Rights, Kathmandu.

Ahmed, I., Dasgupta, A., and Sinha-Kerkhoff, K. (eds). 2004. *State, Society and Displaced People in South Asia*. Dhaka: The University Press Ltd.

Akram-Lodhi, A. Haroon. 2005. What's in a name? Neo-conservative ideology, neoliberalism and globalisation, in *The Neoliberal Revolution: Forging the Market State*, edited by R. Robison. London: Palgrave, 156–72.

Amnesty International Report, 2001. Bangla Hindus: Victims of growing Muslim extremism *Communalism Combat*, 8 (74), 10-17.

Annan, K.A. 2005. *In Larger Freedom: Towards Development, Security and Human Rights For All: Report of The Secretary-General United Nations*. Secretary General New York : United Nations.

Arora, D. 1999. Structural adjustment program and gender concerns in India. *Journal of Contemporary Asia*, 29(3), 328–61.

Banerjee, P. 2002. Dislocating women and making the nation. *Refugee Watch*, 17, December, 6–10.

Banerjee, P. 2005. Resisting erasure: Women IDPs in South Asia, in *Internal Displacement in South Asia*, edited by P. Banerjee, S. Basu Ray Choudhury, and S.K. Das. New Delhi and Thousand Oaks: Sage Publications, 280–315.

Batliwala, S. and Dhanraj, D. 2004. Gender myths that instrumentalise women: A view from the Indian frontline. *IDS Bulletin*, 11-18.

Basu Roy, Arpita. 2002. Afghan women in Iran. *Refugee Watch*, (June), 16–20.

Bayefsky, A and Fitzpatrick, J. (ed.). 2000. *Human Rights and Forced Displacement*, The Hague: Martinus Nijhoff.

Benhabib, S. 2005. Borders, boundaries, and citizenship. *PS: Political Science and Politics*, 38(4, October), 673–77.

Bhaumik, S. 2000. Negotiating access: North east India. *Refugee Survey Quarterly*, 19(2), 142–58.

Bisht, Tulsi Charan. 2009. Development-induced displacement and women: The case of the Tehri dam, India. *The Asia Pacific Journal of Anthropology*, 10(4), 301–17.

Bose, P.K. (ed.). 2000. *Refugees in West Bengal: Institutional Processes and Contested Identities*. Calcutta: Mahanirban Calcutta Research Group.

Boyd, M. 1999. Gender, refugee status and permanent settlement. *Gender Issues*, 17(1), 5–25.

Burkett, P, and Hart-Landsberg, M. 2003. A critique of 'catch-up' theories of development. *Journal of Contemporary Asia*, 33(2), 147–71.

Butalia, U. 1998. *The Other Side of Silence: Voices from the Partition of India*. New Delhi: Penguin Books.

Castles, S. 2003a. Towards a sociology of forced migration and social transformation. *Sociology*, 37 (1), 13–34.

Castles, S. 2003b. The international politics of forced migration. *Development*, 46(3), 11–20.

Cohen, R. and Deng, F. 1998. *Masses in Flight: The Global Crisis of Internal Displacement*. Washington, DC: Brookings Institution Press.

Cohen, R. 2000. 'What's so terrible about rape?' and other attitudes at the United Nations. *SAIS Review*, 20(2), 73–77.

Colson, E. 2003. Forced migration and the anthropological response. *Journal of Refugee Studies*, 16(1), 1–18.

Centre for Women's Development Studies (CWDS). 2000. *Shifting Sands: Women's Lives and Globalization*. Calcutta: Stree.

de Wet, C. (ed.). 2006. *Development-induced Displacement: Problems, Policies and People*. New York: Berghahn Books.

Eastmond, M. 2007. Stories as lived experience: Narratives in forced migration research, *Journal of Refugee Studies*, 20(2), 248-264.

Elson, Diane. 2002. Gender justice, development and rights, in *Gender Justice, Development and Rights*, edited by M. Molyneux and S. Razavi. New York: Oxford University Press, 78–114.

Feldman, S. 2003. Bengali state and nation making: Partition and displacement revisited. *International Social Science Journal*, 55(1), 111–21.

Fukuda-Parr, S. 2006. Millennium Development Goal 8: Indicators for International Human Rights Obligations? *Human Rights Quarterly*, 28, 966–997.

Ganguly, M. 2005. India, after the deluge: India's reconstruction following the 2004 Tsunami, *Human Rights Watch*, 17 (3), 1-47.

Ganguly-Scrase, R. and Julian, R. 1997. The gendering of identity: Minority women in a comparative perspective. *Asian and Pacific Migration Journal*, 6(3–4), 415–38.

Gillan, M. 2002. Refugees or infiltrators? The Bharatiya Janata Party and 'illegal' migration from Bangladesh. *Asian Studies Review*, 26(1), 73–95.

Haines, R. 2003. Gender-related persecution, in *Refugee Protection in International Law: UNHCR's Global Consultations on International Protection*, edited by F. Nicholson, T. Volker, and E. Teller. Cambridge: Cambridge University, 319–52.

Hann, C. 1997. Nation-state, religion and uncivil society: Two perspectives from the periphery. *Daedalus*, (26)2, 27–45.

Hans, A. 2003. Refugee women and children: Need for protection and care, in *Refugees and the State: Practices of Asylum and Care in India 1947–2000*, edited by R. Samaddar. New Delhi: Sage Publications, 355–95.

Harvey, D. 2005. The neoliberal state, in *A Brief History of Neoliberalism*. Oxford: Oxford University Press, 64–86.

Hein, J. 1993. Refugees, immigrants and the state. *Annual Review of Sociology*, 19, 43–60.

Hill, D. 2003. *Policy, Politics and Chronic Poverty: The experience of Bankura District, West Bengal,* Conference Paper, Staying poor: Chronic poverty and development policy, IDPM, University of Manchester, 7–9 April.

Jahid, D. 1996. Islamic fundamentalism in Bangladesh. *Interkulturell*, 4: 44–49.

Kaapanda, M. and Fenn, S. 2000. Dislocated subjects: The story of refugee women. *Refugee Watch*, July, (10-11), 26–29.

Kabeer, N. and Mahmud, S. 2004. Globalization, gender and poverty: Bangladeshi women workers in export and local markets. *Journal of International Development*, 16(1), 93–109.

Kaplinsky, R. 2005. *Globalization, Poverty and Inequality: Between a Rock and a Hard Place*, Polity Press, Cambridge

Kaviraj, S. 2001. The culture of representative democracy, in *Democracy in India*, edited by Niraja Gopal Jayal. Delhi and Oxford: Oxford University Press,

Khan, F. 2005. Gender violence and development discourse in Bangladesh. *International Social Science Journal*, 57(184), 219–30.

Kipnis, A. 2004. Anthropology and the theorisation of citizenship. *The Asia Pacific Journal of Anthropology*, 5(3), 257–78.

Kudaisya, G. and Tai Yong, Tan 2000. Divided landscapes, fragmented identities: East Bengal refugees and their rehabilitation in India 1947–1979, in *The Aftermath of Partition in South Asia*. London: Routledge, 141-162.

Kumar, A. 2010. Illegal Bangladeshi Migration to India: Impact on Internal Security, *Strategic Analysis*, 35:1, 106-119.

Lahiri-Dutt, K. and Samanta, G. 2004. Fleeting land, fleeting people: Bangladeshi women in a charland environment in Lower Bengal, India. *Asian and Pacific Migration Journal*, 13(4), 475–97.

Lindio-McGovern, L. 2007. Women and neoliberal globalization inequities and resistance. *Journal of Developing Societies*, 23(1–2), 285–97.

Lintner, B. 2002. Bangladesh: A cocoon of terror. *Far Eastern Economic Review*, April, p. 4.

Loescher, G. 1999. Refugees: Global human rights and security crisis, in *Human Rights in Global Politics*, edited by T. Dunne and N.J. Wheeler. Melbourne: Cambridge University Press, 233–58.

Loescher, G. 2001. The UNHCR and world politics: State interests vs institutional autonomy. *International Migration Review*, 35(1), 33–64.

Macklin, A. 1995. Refugee women and the imperative of categories. *Human Rights Quarterly*, 17(2), 213–77.

Malkki, L.I. 1995. Refugees and exile: From refugee studies to the national order of things. *Annual Review of Anthropology*, 24, 495–523.

Manchanda, R. 2004. Gender conflict and displacement: Contesting 'Infantilisation' of forced migrant women, *Economic and Political Weekly*, (39) 37, Sep. 11-17, 4179-4186.

McAdam, J. (ed.). 2008. *Forced Migration, Human Rights and Security*. Portland Oregon and Oxford: Hart Publishing.

McNevin, A. 2007. Irregular migrants, neoliberal geographies and spatial frontiers of 'the political.' *Review of International Studies*, 33 (4), 655–74.

Mehta, L. (ed). 2009. *Displaced by development: Confronting marginalisation and gender injustice*, New Delhi: Sage Publications.

Mishra, O. (ed.). 2004. *Forced migration in the South Asian region*. Washington, DC: Brooking Institute.

Nair, R 1997. Refugee protection in South Asia. *Journal of International Affairs*, 51(1), 201–15.

Naruzzaman, M. 2004. Neo liberal economic reforms, the rich and the poor in Bangladesh. *Journal of Contemporary Asia*, 34(1), 33–54.

Niewenhuys, O. 1994. *Children's Lifeworlds: Gender, Welfare and Labour in the Developing World*. New York: Routledge.

Overbeek, Henk. 2002. Neoliberalism and the regulation of global labor mobility. *The Annals of the American Academy of Political and Social Science*, 581(1), 74–90.

Okin, S.M. 2003. Poverty, well-being and gender: Who counts, who's heard? *Philosophy and Public Affairs*, 31(3), 280–316.

Peck, J. 2004. Geography and public policy: Constructions of neo liberalism. *Progress in Human Geography*, 28 (3), 392–405.

Phuong, C. 2005. The Office of the United Nations High Commissioner for Refugees and Internally Displaced Persons, *Refugee Survey Quarterly*, 24 (3), 71-83.

Pittaway, E. and Bartolomei, L. 2002. Refugees, race and gender: The multiple discrimination against refugee women. *Refuge*, 19(6), 21–32.

Qadeem, M. 2003. Internally displaced persons in Afghanistan: A long way home. *Refugee Watch*, 19, April, 17–20.

Rai, S.M. 2002. *Gender and the Political Economy of Development: From Nationalism to Globalization*. Cambridge: Polity Press.

Rajasingham-Senanayake, D. 2004. Between reality and representation: Women's agency in war and post conflict Sri Lanka. *Cultural Dynamics*, 16(2/3), 141–68.

Ramachandran, S. 2002. Operation pushback: State, slums and surrepticious Bangladeshis in New Delhi. *Singapore Journal of Tropical Geography*, 23(3), 311–32.

Razavi, S (ed.). 2002. *Shifting Burdens: Gender and Agrarian Change under Neoliberalism*. Bloomfield, Connecticut: Kumarian Press.

Robinson. G.W. 2003. *Risks and Rights: The Causes, Consequences and Challenges to Development-Induced Displacement*. Washington, DC: The Brookings Institute–SAIS Project on Internal Displacement.

Robison, R (ed). 2005. *The Neoliberal Revolution: Forging the Market State*, London: Palgrave.

Rodriguez Bello, C. 2003. *Refugees and Internally Displaced, WHRnet*. (Available at: http://www.onlinewomeninpolitics.org/beijing12/03_10_women_media.htm [accessed: 23/09/2011].

Rudolph, C. 2005. Sovereignty and territorial borders in a global age. *International Studies Review*, 7(1), 1–20.

Rudra, Nita. 2002. Globalization and the decline of the welfare state in less-developed countries. *International Organization*, 56(2, Spring), 411–45.

Samaddar, R. 1999. *The Marginal Nation: Transborder Migration from Bangladesh to West Bengal*. New Delhi: Sage Publications.
Samaddar, R (ed.). 2003. *Refugees and the State: Practices of Asylum and Care in India 1947–2000*. New Delhi: Sage Publications.
Singh, P. 2002. Demographic movements: The threat to India's economy and security. *Low Intensity Conflict and Law Enforcement*, 11(1), 94.
Singh, S .2009. "Border Crossings and Islamic Terrorists": Representing Bangladesh in Indian Foreign Policy During the BJP Era, *India Review*, 8 (2), 144-162
Sivanandan A. 2001. UK: Refugees from globalism. *Race and Class*, 42(3), 87–100.
Sorensen, B. 2003. Anthropological research on forced migration: Contributions and challenges. *Acta Geographica*, 6, 65–83.
Steger, M.B. 2005. Ideologies of globalisation. *Journal of Political Ideologies*, 10(1), 11–30.
Stevens, D. 2003. Roma Asylum Applicants in the United Kingdom: "Scroungers" or "Scapegoats" ', in *The Refugee Convention at Fifty: A View from Forced Migrations Studies*, edited by Joanne van Selm et al., Maryland: Lexington Books, 145–60.
Stravropoulou, M. 1998. Displacement and human rights: Reflections on UN practice. *Human Rights Quarterly*, 20(3), 515–54.
Suhrke, A. and Newland, K. 2001. UNHCR: Uphill into the future. *The International Migration Review*, 35(1), 284–302.
Sunder Rajan, R. 2000. Women between community and state: Some implications of the Uniform Civil Code debates in India. *Social Text*, 18(4), 55–82.
Upadhyay, U.D. 2000. India's new economic policy of 1991 and its impact on women's poverty and AIDS. *Feminist Economics*, 6(3, 1 November), 105–22.
Van Schendel, W. 2003. Working through partition: Making a living in the Bengal borderlands, in *Work and Social Change in Asia, Essays in Honour of Jan Breman*, edited by A. Das and M. van der Linden. New Delhi: Manohar, 55–89.
Vervoorn, A. 2006. *Reorient: Change in Asian Societies*. Melbourne: Oxford University Press.
Waas, J., van der Kwaak, A., and Bloem, M. 2003. Psychotrauma in Moluccan refugees in Indonesia. *Disaster Prevention and Management*, 12(4), 328–35.
Weiner, M. 1996. Ethics, national sovereignty and the control of immigration. *International Migration Review*, 30(1), 171–97.

Chapter 5
Aftermath of Dams and Displacement in India's Narmada River Valley: Linking Compensation Policies with Experiences of Resettlement

Pratyusha Basu

Introduction

The connection between economic development and community-level displacements has become a key theme in analysing processes of contemporary globalisation, with many scholars arguing that displacements have become markedly more brutal due to the worldwide recessionary climate and the need for new spaces and resources to extend corporate profitability (Moore 2008; Harvey 2003). Yet, to the extent that the scale of global economic development concentrates on transnational capital flows and resistance networks, locally differentiated experiences of displacement within national contexts are not drawing an equivalent amount of attention. This chapter aims to redress such absences through engaging with the aftermath of displacements due to the Narmada river dams in India. The social movement mobilised against the construction of the Narmada dams has questioned the extent to which existing compensatory frameworks actually enable the displaced to rebuild their lives in new locations. This, however, has not prevented the central and state governments in India from persisting with dam construction and their associated displacements. Given that development-related displacements will continue to characterise India as well as the broader Global South, attentiveness to post-displacement experiences becomes an especially crucial approach to building complex and critical perspectives on the nature of contemporary mobility.

The debates around displacement due to development are well brought out in *A Narmada Diary*, Anand Patwardhan's 1995 documentary on the people's movement against the Narmada valley dams. In this documentary, a comment by a Gujarati man opposing a rally led by the anti-dam Narmada Bachao Andolan (Movement to Save the Narmada River) is very illuminating because of the comparison it presents between displacement due to development and women's displacement due to patrilocal traditions (Patwardhan and Dhuru 1995). Likening the displaced to newly married women, he suggests that they will lament the loss of their original homes for a few days, but with the passage of time will soon adjust to their new situation. The sorrow of displacement is thus represented as transient, as momentary as the lament of a woman who has just arrived at her husband's place. I brought up this comment during an interview with a prominent activist, a young woman, associated at that time with the Andolan's struggle against the Maheshwar dam. She did not take kindly to the analogy. According to her, marriage provides women the opportunity to build new relationships, to extend their range of identities by becoming mothers, wives, and daughters-in-law. Displacement due to development in contrast destroys existing social networks and does not provide access to new ones.

This chapter shows how post-displacement communities are formed through interactions between government policies, social divisions, and place-based attachments. More specifically, it seeks to link modes of compensation to experiences of resettlement. Focusing on cash-for-land, land-for-land, and rights to the dam reservoir as three different ways in which the Narmada dam displaced have been or can potentially be compensated, this chapter shows how different kinds of post-displacement subjects are constructed through each compensation policy, further intensifying already existing gender, class, and cultural divisions. Cash compensation assumes an ability to successfully integrate into the market economy on the part of marginalized social groups. Resettling populations close to their original places of residence around the dam site is a third mode of compensation that should be considered alongside cash and land-based compensation.

Dams on the Narmada River in central India provide valuable insights into displacement since a range of compensatory techniques can be explored through them, partly due to the large-scale nature of this valley-wide project. The Narmada dam-related resettlements also enable a reflection on how social inequalities, and misunderstandings of rural livelihoods in general, are accentuated through such compensations, shedding light on how hegemonic notions of the proper rural subject and national citizen are imposed through displacement. This chapter draws partly on existing studies of displacement and resettlement in the Narmada valley, and partly on my own observations, ongoing since 1999, of displacement and resettlement linked to the Sardar Sarovar dam in the state of Gujarat, and the Maheshwar, Indira Sagar, Maan and Tawa dams in Madhya Pradesh. In the process, this chapter makes two main contributions to existing studies of displacement. First, by drawing on materials that directly engage with the perspectives of displaced communities, this chapter provides a range of in-depth, experiential accounts. These serve as valuable counterpoints to official discourses, which, at the very least, suggest equivalence between what is lost through displacement and what is regained through resettlement, and, at the extreme, argue that the benefits of displacement far outweigh its costs. Second, while state-based compensatory frameworks are often geared towards uniform solutions for displacement, this chapter's attentiveness towards social and cultural inequalities within displaced communities shows how fractures within nation states are actually heightened through existing compensation policies, leaving local communities with the burden of bridging gaps between place-based identities and imposed mobilities.

At the outset, the close relationship between development and displacement is outlined. While rural populations are increasingly bearing the brunt of displacement, the supposed benefits of development often lead to an undercounting of these costs. The anti-dam Narmada movement is then introduced in terms of its specific demands related to resettlement and wider opposition to the violence unleashed on rural and tribal communities by existing forms of economic development. Then, the three main kinds of compensations that are available to the Narmada-dam displaced – cash compensation, land-for-land, and rights to the dam reservoir – are examined. In the process, the struggles that had to be undertaken before the need for compensation was officially accepted by the government, how each kind of compensation can be linked to particular experiences of resettlement, and the extent to which such compensation actually enables successful rehabilitation of the displaced are considered. Overall, this chapter seeks to argue that experiences of resettlement are linked not only to the specific reasons for displacement and the characteristics of displaced communities, but also to the resources that the displaced bring with them as they are forced to seek new lives.

Development and Displacement

The significance attached to mobility in the pursuit of development is highlighted in a recent United Nations's *Human Development Report* (2009). Positing that international migration raises a host of contentious issues, the report seeks to exclusively focus on the benefits of internal migration as a policy to redistribute development. However, this does not address either the social justice implications of making participation in development conditional on mobility for economically vulnerable sections of the population, nor does it take into account the extent to which causes of underdevelopment are often linked to the global economy and cannot be corrected within national contexts (Raghuram 2009). Given that ability to partake in the economic benefits of migration usually requires specialised education and training, the experiences of relatively privileged classes is in danger of being utilised as paradigmatic of migration outcomes in general. Moreover, the assumption that rural-to-urban migrations are part of the natural pursuit of development needs to be carefully examined since the extent to which participation in the urban cash economy is either economically more secure or socially more valuable than subsistence rural livelihoods is uncertain (United Nations 2009).

Internal displacements become a very useful lens through which to expand existing ideas about the forms and values of mobility, enabling the argument that economic globalisation-led migration is not merely experienced in terms of a crossing of international borders (Skeldon 2006). As displacements are becoming a key feature of contemporary development, scholars have begun to view recent dislocations as yet another round of primitive accumulation. Harvey (2003) focused on the geographies of contemporary economic policies to show that rural areas are facing the consequences of a new round of capital accumulation under the aegis of neo-liberalisation, a process of privatization of resources that Harvey designates as 'accumulation by dispossession'. A turn towards foreign investment flows as a means to fund national development means that the bulk of development-induced displacements are in the interests of capturing natural resources for multinational capital. Internal migrations caused by loss of access to natural resources and encroachments on traditionally held land become a perverse form of participation in globalisation – as capital seeks relatively remote rural resources to intensify its powers of accumulation, local inhabitants are encouraged to become mobile without any assurance that such mobility will enable their entry into either national or global economies. The value of migration for understanding the micro-processes of capitalist development has been highlighted by Lawson (2000). For her, the theoretical potential of the experiences of poor, rural migrants is linked to their location at 'the margins of the urban economy' (Lawson 2000: 186). Displacement due to dams can be counted as similar sites of clarified understanding of mainstream development through an appreciation of those who are marginalised and rendered mobile by it.

Recent studies have begun to emphasise the need to build an interconnected understanding of mobility and immobility. Yeoh et al. (2003) have used the notion of 'transnational edges' to conceptualise how places are comprised of differential forms of mobility. A focus 'on the "edges" of transnationality, [shows that] what constitutes "groundings" which locate people in particular places as opposed to "unmoorings" which destabilize these localizations have become inextricably linked' (Yeoh et al. 2003: 208). Lubkemann's (2008) study of war-related violence in Mozambique shows how displacement can occur without any movement, as landscapes changed by violence become new places for those who continue to inhabit them. From this, Lubkemann argues for a need to expand the meanings of mobility in order to incorporate those forms which do not comprise any actual movement. In his words,

> .... preconceived notions about the relationship between mobility and social place tend to render invisible an entire category of people whose lives are as profoundly (and sometimes more) disturbed by the effects of conflict on the mobility environment as are the lives of wartime migrants. However, this category of the 'displaced' ultimately move little if at all. (Lubkemann 2008: 456)

It could be argued that the changes wrought by dams also produce unfamiliar places without any movement on the part of residents.

Dams become a poignant case of displacement because they destroy the possibility of return to an original home, since the home itself no longer exists. The specific trauma of this loss is depicted vividly in Read's (1996) discussion of the aftermath of dam building in Australia. Read focuses on the stages through which two towns in New South Wales were evacuated for eventual submergence due to dams, concluding his narrative at a point when an unusual decline in water levels has caused a previously submerged set of houses to reappear. As people returned to look at their houses, it became clear that connections to their dwellings had not been lost and this momentary uncovering revealed the losses that are embedded in development projects. Such losses, however, cannot be properly mourned due to the ubiquity of the notion that dams are contributions to national progress, making any lamentations of the loss of homes an unpatriotic act. In the context of the Narmada dams, Roy (1999) has discussed the tyranny of the 'greater common good' argument made in support of dams, which exaggerates benefits and prevents a cogent counting of losses due to dams.

A focus on displacement is significant for understanding both dominant forms of economic development and social groups which are marginalised in the process. Resettlement experiences provide a way in which to approach displacement as a long-term process, unfolding beyond the immediate experience of loss of place due to development (Fujikura et al. 2009). The Narmada dams are a valuable case study of displacement and resettlement since their geographic range encompasses not only a diversity of social groups within India, but also a series of compensation policies which provide a means to examine the contextual specificities of the aftermath of displacement.

Dams and Displacement along the Narmada River

The social justice implications of displacement due to development have become the subject of intense debate in India partly due to the endeavors of the Narmada Bachao Andolan, a social movement that opposes the large-scale human displacements caused due to the Narmada river dams in western India (Ramachandra 2006). The Andolan has been adept at the construction of transnational alliances so that a more global questioning of and resistance against displacements due to dams, coordinated in part by activist organisations like International Rivers in Berkeley, California, and reflected in the establishment of the World Commission on Dams, has now emerged. The movement became especially celebrated because of its role in forcing the withdrawal of World Bank funding for the Sardar Sarovar dam in 1993. Multinational dam construction and funding entities, however, have continued their projects in India and across Asia, Africa, and Latin America by providing assurances of greater sensitivity to the interests of displaced populations and natural environments (for example, Ford 2009). The aftermath of dam building, specifically in terms of modes of compensation for displaced populations, thus continues to be a matter of concern (Parasuraman et al. 2009).

The history of dam building and dam-related displacements in India reflects both colonial legacies (Gilmartin 1995) as well as the emphasis on large-scale agricultural and industrial modernisation under the rubric of national development (D'Souza 2006, Klingensmith 2003). Initial plans related to the building of dams on Narmada can be traced back as far as 1946, to the eve of India's independence from colonial rule, so that dam-building projects in India suggest continuity between colonial and postcolonial desires for development. The Narmada Valley Development Project envisages the construction of 30 large dams, 135 medium dams, and 3,000 small dams on Narmada and its tributaries, and this attempt to harness an entire river valley is an extreme representation of the large-scale control of nature that is a key aspect of contemporary development. The Sardar Sarovar, which is the largest dam on Narmada, began in 1961 as a small dam dedicated to flood control, and now more grandly aims to direct the flow of Narmada from south to north Gujarat and supply drinking and irrigation water to the Kutch region and parts of Rajasthan. According to Worster (1985), such large-scale initiatives are productive of imperialist relationships since their control and management require correspondingly large institutions which centralise power and become inherently non-democratic.

The politics of the Narmada Valley Development Project are further complicated by the fact that four states – Gujarat, Madhya Pradesh, Maharashtra, and Rajasthan – were part of negotiations for the sharing of its waters. While Narmada does not flow through Rajasthan, this state is slated to receive benefits from the project. The inclusion of arid Rajasthan can be partly viewed as an attempt to enhance the significance of the Narmada dams. However, the possibility of water reaching north Gujarat and Rajasthan has been called into question, since engineering, environmental, and political factors might intervene to thwart this transfer. The issue of borders emerges most prominently in the case of the Sardar Sarovar dam, since the benefits of the dam will flow mostly to the state of Gujarat, while the bulk of the submergence is occurring in Madhya Pradesh. Not surprisingly, electoral politics in Gujarat are constructed around unequivocal support for the Sardar Sarovar, while the official position on the dam is more contradictory within Madhya Pradesh. While there has been some attempt to resettle dam oustees from Madhya Pradesh in the command area of the Sardar Sarovar dam in Gujarat, a wholesale transfer of the displaced from one state to another is neither economically nor culturally feasible.

The Narmada Bachao Andolan has questioned the link between dams and economic development given the continued marginalisation of tribal and rural populations due to dam-related displacements in India (Basu 2010). The movement's origins can be traced to the beginnings of World Bank funding for the Sardar Sarovar dam in 1985. As scholars and activists began investigating the Sardar Sarovar, a number of them were aghast at how little was known of the consequences of the dam by people who would be directly affected by its construction (MARG 1991). This led to the formation of organisations seeking to ensure that the rights to compensation of the displaced were not violated. In 1989, however, a rift occurred between those organisations which sought to continue the struggle for better compensation and those who had become certain that the government had neither the resources nor the will to provide adequate compensation to the displaced. The Andolan was formed from this latter point of view and included organisations that had decided to move from seeking better resettlement to unequivocally opposing the construction of dams, especially large-scale dams. In the process, Andolan activists documented the multiplicity of traditional connections to land and water that could not be justly compensated. Moreover, they also pointed out historical experiences of multiple displacements that characterised rural places, so that sacrifices in the name of development were constantly being sought from those who had remained outside the benefits of development. In Nielsen's (2008) formulation, the Andolan's arguments demonstrated the breakdown of the 'postcolonial compact' – the state now could no longer be viewed as a guarantor of the rights of the weakest sections of the nation.

The process of land acquisition for development projects is widely viewed to be in need of overhaul since the Land Acquisition Act, 1894, was originally drafted for the purposes of colonial rule. As Maitra (2009: 198) points out:

> Appropriate land for the development projects was acquired by the state under the 'Land Acquisition Act' (1894). Subject to subsequent periodic amendments (with substantial amendments being made in 1984), the government of independent India retained this colonial legal instrument to establish the principle of 'eminent domain'. Thus, the Act justifies the takeover of land by the state for 'public purpose' under the principle of eminent domain.

Studies have been critical of this vague notion of 'public purpose,' pointing to its misuse in favour of implementing development projects without reference to existing uses of the space and resources sought to be developed (Parasuraman et al. 2009).

In 1993, the Andolan won a major victory as the World Bank withdrew funding from the Sardar Sarovar dam. However, the Gujarat state government immediately announced that it no longer needed international funding for the dam, so that the Andolan now had to shift to a new tactic of resistance – choosing to fight the state and central governments in the courts. By 2000, however, the precariousness of the legal option became clear as the Andolan's case against the Sardar Sarovar dam was rejected by the Supreme Court. Ample evidence of the failure to resettle the displaced could not stymie dam construction, and one of the consequences of this has been the Andolan's more overt entry into the struggle for better compensation policies.

The number of people already displaced as well as facing displacement due to the Narmada dams and its canals is difficult to estimate. Figures put forward by the government have been challenged by the Andolan as underestimations both because of inaccuracies in village surveys as well as due to the government's consistent refusal to accommodate various forms of displacement that will be unleashed by changes in the river. But it is not just problems associated with large numbers or means of identifying the displaced that become troubling in terms of determining compensation. Inequalities that comprise Indian society and the possibility of further intensification of class, caste, and gender divides through displacement make compensation a complicated matter. Needs of farmers dependent on agriculture are likely to be different from fisher communities dependent mainly on the river or tribal communities accustomed to supplementing everyday lives through access to forests.

Existing social and economic differences are exacerbated by the form of compensation provided to the displaced. It is for this reason that the Andolan favours land-based compensation over cash compensation, the latter being viewed as leading to eventual paupersiation. More innovatively, the Andolan has also sought to ensure that the displaced are provided rights to the reservoir of dams, adding a new form of compensation to supplement land-based policies. In the remainder of the chapter, I address the extent to which existing compensation policies can be directly linked to the production of floating and fragmented communities, even as local communities continue to struggle to remake place for themselves.

Cash Compensations and Floating Populations

The option to obtain compensation in the form of cash is opposed by the Andolan as well as the World Bank, and both emphasise the need for land-based forms of compensation. Yet, cash continues to be one of the ways in which the displaced are sought to be compensated, and this is especially true in the case of the Madhya Pradesh government (Maitra 2009). The lack of suitable agricultural land is one reason given for the resort to cash compensation, but governments have also argued that the displaced themselves often express a preference for cash compensation. This ignores the manifold problems associated with distribution of cash compensation, including the danger of corruption in the form of middlemen and officials who could expect a bribe in return for providing compensation, and the greater power of wealthier families to be able to manipulate local government officials and gain access to cash.[1] Within villages, the unevenness of cash compensation results in a further breakdown of community networks, as large farmers are more likely to be able to gain access to cash compensation than landless labourers.

Another problem with cash compensation is that it is not accompanied with any training in viable ways of spending, investing, or saving cash. For those who have remained somewhat distant from the cash economy, sudden access to a large amount of money is perplexing due to lack of suitable norms for its utilisation. The value of the provision of cash compensation thereby has to be considered with reference to the ways in which that cash will be spent. If investment is sought to be made in the procurement of agricultural land, escalation of land prices in the face of increasing demand often makes it difficult to purchase an adequate amount of land. In most cases, there is a propensity to utilise cash for forms of consumption that are not productive of long-term livelihoods. Tales abound of displaced families who have spent their cash on buying luxury consumer goods

1 Personal communication in village facing submergence due to the Sardar Sarovar dam near Badwani town, Madhya Pradesh, May 2002.

(for example, automobiles) or marriage expenses, and now have been reduced to the status of labourers.[2] Similar findings have been reported by Swainson and McGregor (2008: 164) in the case of the Sungai Selangor dam in Malaysia where '[c]ash compensation has led to increased purchases of consumer goods but not contributed to any sustained livelihood opportunities...' In the case of the Narmada dams, cash compensation has usually been provided to male heads of households with women being automatically considered as dependents, as a result removing women from decision-making over spending the compensation amount (Thukral 1996). Among some displaced families, cash compensation has intensified alcoholism among men or enabled them to choose a new marriage partner often deserting their previous families.

In the long-term, cash compensation requires an entry into the job market to maintain access to income. But the absence of jobs that would utilise existing agricultural skills and the lack of training for industrial or service-related employment means that the displaced are not always able to obtain adequately paid or secure jobs. The tendency often is to migrate from rural to urban areas in search of jobs. Some displaced families become part of slum populations adding to already burdened informal urban economies. Where migration is between rural areas, the entry of new migrants is likely to increase competition among agricultural labourers. Differences between the more-urbanised Gujarat economy versus the less-urbanised Madhya Pradesh economy are also likely to impinge on the ability of the displaced to fit into new forms of livelihood. Gendered aspects of access to jobs become an important consideration here. In a study of women's access to jobs after migration, Banerjee and Raju (2009) have found that women previously engaged in agricultural work are likely to move to manufacturing work after rural-to-urban migration. Their study showed that illiterate and higher-caste women were less likely to find employment compared to lower-caste women. The link between the social characteristics of the displaced and existing job markets becomes a key aspect of ability to reintegrate into new contexts after displacement. It has been argued that the displaced should be provided jobs in construction at the dam site on a priority basis and contracts for construction should be signed with this understanding. Asking the dam-affected to help in the very project that displaced them verges on an act of cruelty. Moreover, the lack of long-term job security as well as the lack of adequate remuneration casts doubts on the value of such construction jobs.

Noteworthy about the provision of cash compensation is that it becomes a means for the displaced to disappear from rural to informal urban economies and ultimately from most forms of official accounting. In other words, cash compensation produces floating, disconnected populations that serve to further hide the consequences of displacement. The provision of land as compensation provides a target group within which post-displacement experiences can be more

---

2 Personal communication from Maan dam oustees in Dhar district, Madhya Pradesh, June 2002.

carefully documented. The following section seeks to document the consequences of land-based forms of compensation through focusing on resettlement sites and the provision of agricultural land.

Land for Land: The Struggle to Replace Homes and Livelihoods

Land-based compensation is the centrepiece of the struggle of the Andolan, and this strategy has also been stipulated by the World Bank as one most likely to rehabilitate displaced populations. Ideally, rehabilitation should initially reinstate displaced populations to their original standard of living and ultimately improve their living conditions. Resettlement policies of the government of India provide land-based compensations through resettlement sites and associated agricultural lands. This form of compensation continues to be dogged by problems because of shortages of agricultural land and due to the fact that the forms of communities and livelihoods that are being destroyed cannot simply be recreated in new places. The experiences of resettlement associated with the Narmada valley dams raise concerns related to the ways in which changes in economic and social identities are experienced by the displaced and the extent to which government policies have failed to account for such changes. A careful understanding of the social networks and material practices left behind in places of origin and the processes of settling into new places is thus required.

Resettlement Sites

The Andolan has emphasised that existing communities need to be resettled as one unit in order to maintain social cohesiveness after displacement. This has been a difficult demand to implement partly because of the lack of sufficient land to ensure that both homes and fields can be provided for all households in one site, and partly because the insertion of the displaced among new communities already ensures that previous social networks cannot be maintained. One issue of particular concern is the emergence of conflicts between existing residents and new arrivals. Where tribals have been resettled within predominantly non-tribal contexts, social tensions are likely to be further intensified. Where displaced communities from Madhya Pradesh and Maharashtra have had to move into resettlement sites in Gujarat, this has entailed a move into a different language area (TISS 1997). Existing prejudices against tribals come to the fore in such conditions. Social tensions are reflected in not being able to fully utilise agricultural lands in the absence of cooperation from existing farmers, or in problems with finding employment as agricultural labourers. The extent to which the possibility of conflicts due to cultural differences or economic competition has been incorporated in the allotment of resettlement sites is not clear, since land is likely to have been acquired by the government wherever it becomes available.

The quality of infrastructure, including electricity and water connections, schools and playgrounds, have usually been represented as marking an improvement over the previous living conditions of the displaced. This representation, however, could be inaccurate in two ways. To begin with, resettlement sites are built with the assumption that previously class divided communities will now be willing to inhabit similar forms of housing. Then, water, electricity, schools, and health care facilities at resettlement sites are usually linked to the regular payment of bills, and in the absence of steady employment or income, it is not clear if displaced families can take advantage of the infrastructural offerings of resettlement sites (Hakim 1997: 155). Similar findings have been reported by Swainson and McGregor (2008: 162) in their study of resettlement associated with the Sungai Selangor dam in Malaysia. They point out that the inability to afford new utilities led to social embarrassment further adding to the discomfort of resettlement.

One main concern raised about the quality of the built environment in Sardar Sarovar resettlement sites is related to the lack of protection against heavy rains which has often led to floods and waterlogging.[3] Moreover, houses in rural areas are often built from natural materials, and the concrete and tin used in resettlement sites do not protect inhabitants against heat. While displaced populations can take the materials of their original houses with them to resettlement sites, this is often rendered impossible due to inability to pay transportation costs.

A significant aspect of resettlement is the loss of place-based identities. This becomes pressing in the case of tribal connections to land, but is also more generally true of rural cultures based on attachments to specific environments. According to Baviskar's study of Bhilala communities facing submergence due to the Sardar Sarovar dam, the '*gayana*', a creation myth of this community, links it to a distinctive place. In Baviskar's (1997: 123) words, '[t]he religious conjunction of the natural and social worlds is marked cognitively such that the village is defined with respect to a specific site and to its ancestors who inhabit the site'. Xaxa (1999), in her study of the Vasava community, also showed how displacement truncates existing tribal identities, producing new forms of collective belonging which reduce social spaces available to displaced communities. According to Xaxa (1999: 1494),

> Vasava self-identity is articulated in terms of their economic self-sufficiency in production and consumption. The community is very proud of their economic independence and often use this to articulate their difference from other groups. If [however] … resettled Vasavas begin to use consumption and religion as a basis for their identity instead of self-sufficiency in production and consumption (as used by hill Vasavas), this will affect inter-Vasava relations at both the individual and collective level.

3 Personal communication from resettled families in a Sardar Sarovar resettlement site in Gujarat, January 2000.

This is not to suggest that tribal identity is static and would not show changes in the absence of displacement, but that the creation of new identities after resettlement could lead to a lack of integration with both old and new community networks.

Women are most likely to be affected by such loss of community networks. To begin with, living in resettlement sites is likely to raise issues of personal security, as women travel to nearby communities as part of their domestic duties or for employment (Mehta and Srinivasan 2000). Given that gendered practices within tribal communities often differ from caste Hindu norms, tribal women face greater harassment. Then, due to norms of patrilocality, women often travel between natal and marital homes; such travel becomes difficult when resettlement sites are located away from natal homes. The loss of support from natal homes is likely to render women even more insecure after displacement (Basu and Silliman 2000: 427–29). In a study of Sardar Sarovar resettlement sites, Chattopadhyay (2009: 13) has found that:

> ... women are less mobile in terms of wage labour and market accessibility than their male counterparts ... [which] reflects the social construction of the patriarchal society's notions of gender, work and mobility, demarcating their movement within the private spaces – household, family farm, and the village.

Yet, Chattopadhyay also points out that younger women are seeking to change the constraints within which their lives are lived through access to education and the delaying of marriage. Breakdown of traditional social norms in resettlement sites provides a sense of freedom from oppressive gendered practices in this case.

Resettlement sites do not automatically produce community, and it is in their link to livelihoods that the sustainability of resettlement should be examined. Residential spaces and agricultural land have not always been offered in tandem, so that issues with resettlement also emerge in the disjunction between provision of housing and access to income-earning opportunities.

Agriculture Out of Place

A key stipulation of the land-for-land compensation framework is that both land-owning and landless families are entitled to land. This is a progressive feature of current resettlement policies, but its actual implementation has been marred both due to an absence of suitable agricultural land as well as limited notions of eligibility for land ownership. Land titles are usually reserved for men, which puts women in general and women-headed households in particular in a highly precarious position. Moreover, adult sons and daughters often find themselves missing from lists of beneficiaries. Problems with accurate surveys of displaced communities continues to be a problematic feature of compensation, with certain neighbourhoods often left out of surveys due to inaccurate understandings of village boundaries or deliberate misrepresentations on the part of authorities.

Obtaining agricultural land in itself is often not sufficient to enable resettled households to install a viable agricultural operation. According to Parasuraman (1997: 61), one policy aim was to '[provide] resettlement in command areas which will eventually have the potential for wage labor and self-employment'. Yet, such land is unlikely to be relinquished by existing farmers, so that acquiring irrigated land for displaced families becomes difficult, and agricultural land that becomes available to displaced populations is often of low quality. Further, as the Center for Social Studies (1997: 228) points out, there is often a lack of clear title to land which means that improvements to land become less feasible. An additional point to consider is that the supplementing of agriculture through access to community forests and grazing land is often not possible in resettlement sites, especially in sites of intensive agriculture.

Agricultural work is organised around new forms of labour arrangements in sites of resettlement and this has repercussions on the viability of agriculture. Hakim (1997: 148) points out that displacement leads to a lack of access to kin who can provide agricultural labour. Kedia (2004: 433) finds similar decrease in access to kin labour among communities displaced by the Tehri dam in north India. Mehta and Srinivasan (2000) survey existing studies on resettlement associated with the Sardar Sarovar dam to argue that greater demands on women's time for agricultural work results in decreased mobility for women. Moreover, unfamiliarity with commercial agriculture as well as increasing mechanisation of agricultural work can lead to the exclusion of women from household decision-making in which they had previously been significant participants. For agricultural labourers, resettlement is likely to lead to decreased work as they have to compete with already present labourers, which in turn could further drive down prevalent wage rates.

Agricultural practices are not easily portable so that differences in the kinds of crops cultivated are likely to require new skills in fieldwork and marketing. As Chattopadhyay's (2009: 8) study among Sardar Sarovar oustees belonging to tribal communities argued:

> … [v]illagers who are ineffectual in using advanced farming methods are falling behind those who have acquired considerable expertise in obtaining high productivity and speculating in crop and credit markets. In addition, those who received infertile and uneven land as compensation are falling into the debt trap.

Hakim's (1997: 145) study pointed out dissonances between rhythms of agriculture as followed by tribal communities, marked by particular festivals, which may not be in accordance with the more intensive cycles of commercial agricultural production prevalent in sites of resettlement.

Changes in cropping patterns have implications in terms of diets so that changes in crops grown and shift from subsistence to commercial production could impinge on nutritional options. Chattopadhyay (2009: 8) attests to this possibility; an interview respondent points out the following:

> Previously, we supplemented our diets with a wide variety of vegetables grown on river banks; fish from the river; a variety of fruits, nuts, leaves and tubers from the forests; and milk from cattle... The pressure to cultivate cash crops is high for the increased farm and household expenditure. Therefore, now we have limited possibilities to cultivate food crops.

Kedia found similar consequences after displacement due to the construction of the Tehri dam in north India. He found that diets changed 'from the locally produced, diversified, high-protein diet they maintained while living in the mountains to a nutrient-poor, high carbohydrate diet, primarily based on what was available and affordable in the market' (Kedia 2004: 439). Lack of access to suitable agriculture and transformation of traditional agricultural practices undermines the long-term health of displaced communities.

A vivid description of the problems with resettlement is provided by Whitehead in the context of the Sardar Sarovar dam. In her words (2000: 3976):

> It was pure shock at the difference between the impression I received from reading the reports and what I personally witnessed from visiting resettlement sites in supposedly one of the most favourably irrigated and best situated resettlement regions. Complete crop failure due to waterlogging, death and/or distress sales of livestock, sale of women's silver, distress sales of teak and 'kheir' house posts and furniture, one or several members of every household going to the roadway to search for daily agricultural work from neighbouring farmers, young adult sons without land or employment and drinking heavily, tin shacks leaking water badly, households forgoing a midday meal to decrease their consumption costs, permanent migration to Bharuch of some for construction work, people trying, but being stopped, from returning to submerging villages, and even two resettlement sites that were nearly deserted. The reports that I previously read at the Centre hardly prepared me for these conditions.

The conditions Whitehead reported for 1997–1998 are partly corroborated by Chattopadyay's 2004 survey, and it is hoped that continuing surveys will further address the complexity of resettlement experiences. The broader picture that has emerged so far is that the provision of agricultural land without the social relations necessary to cultivate traditional crops and gain access to labour becomes equivalent to loss of agricultural livelihoods. Moreover, since rural livelihoods are often constituted at the nexus of land, water and forests, the provision of land by itself constitutes a form of environmental deprivation.

The studies conducted on the Narmada dams and on other dams suggest that the post-displacement period is one of trauma for oustees and that even land-based compensation leaves much to be desired. Those displaced by development are thus ultimately excluded from the benefits of development. The next section brings up one more form of compensation that can be used to supplement land-based compensation by recounting how rights to reservoirs can add to the rehabilitation of displaced communities.

Ownership of Dam Reservoir: From Land to Water-based Livelihoods

While land-for-land remains a desired principle of rehabilitation, the supplementing of agricultural livelihoods with water-based livelihoods, like fishing, also needs to be considered to extend the range of compensation. In the case of the Narmada dams, rights to the reservoir have become an important part of the movement's demands on behalf of oustees; in case of the Bargi and Tawa dams, such rights have begun to be made available to oustees. The Tawa dam in Hoshangabad district, Madhya Pradesh, is a markedly tragic experience of dam construction. Completed in the 1970s, before the anti-dam movement became prominent in India, those displaced by the Tawa dam were not provided with compensation. Communities which lost land resettled themselves around the reservoir. The oustees suffered further displacements due to a military firing range and the establishment of a protected forest in the region. The Korku and Gond communities around the dam were initially designated as poachers when they sought to utilise the resources of the reservoir. Subsequently organised by the Kisan Adivasi Sangh (Farmer–Tribal Union), the Tawa dam displaced began a struggle for rights to the reservoir. They were successful in forming the Tawa Matsya Sangh (Tawa Fishing Union) in 1996 (Vikas 2001; Sunil and Smitha 1996). This consists of village-level fishing cooperatives with the Sangh responsible for collecting, packing, transporting, and marketing the fish. The location of the Tawa reservoir near Khandwa, a small town that is a prominent hub of India's railway network, has facilitated access to urban markets.

In the case of the Tawa dam oustees resettlement options have been sought around the site of their displacement. This cannot be considered an absence of displacement as the landscape around them has been completely transformed. Is it a form of trauma to have to live in sight of their submerged homes? Or does the ability to retain tangible memories of displacement actually ensure a basis for the continuity of community identity? As reported by Swainson and McGregor (2008: 164) in their study of resettlement associated with the Sungai Selangor dam in Malaysia:

> [t]he Peretak community [which] continues to access its traditional lifestyles, homelands and environments ... are generally satisfied with their new village location and facilities. Their levels of satisfaction differ markedly from that of Gerachi residents who have received similar social and economic compensation but had to sacrifice easy access to their traditional lands.

However, Duflo and Pande (2007) show that districts in which dams are constructed tend to become poorer over time. Given that rights to the reservoir would compel displaced communities to stay in place, the extent to which these rights would help overcome the possibility of long-term impoverishment has to be considered.

One of the biggest obstacles to providing reservoir rights to the displaced are current moves to privatize dam reservoirs in the interests of commercial fishing or tourism ventures . Access to the reservoir needs political mobilisation on the part of displaced communities, and the experience of the Tawa Matsya Sangh has shown that the struggle to maintain fishing rights has to be continuous. There is a need to formalise access to the reservoir as part of compensation for displacement due to dams. While this does not mitigate the consequences of displacement, it would signify a broader commitment to the rights of the displaced.

Conclusion

Dams have been under construction in the Narmada valley since the 1960s; such construction is reaching its peak currently as a number of large dams are reaching their full height. It is imperative in this context to retain a focus on displacement, and this chapter has sought to do this through incorporating the viewpoints of people displaced by various dams. In all cases, powerlessness and inequalities have been accentuated due to displacement, underlining the need for a national policy to ensure that development does not leave a trail of destruction in its wake.

Based on the discussions above, some conclusions can be drawn on the differences between various forms of compensation. Land and reservoir-related compensations, to the extent that they enable access to and some control over material landscapes and, hence, to specific territories, are relatively more conducive to maintaining the rights of the displaced. Cash compensation is not linked to specific places: it produces floating populations that cannot be tracked and their futures become invisible in accountings of development. As Franke (2008) has pointed out, human rights are usually linked to specific national contexts, and migrants who cross national borders, especially illegal immigrants and refugees, become less amenable to being accommodated within existing understandings of human rights. In his words, '[i]nsofar as the subject of territorially defined civil society remains the central subject of contemporary human rights discourse, displaced persons, as such, at best can only be secondarily human in their status and claims' (Franke 2008: 275–76). Cash compensation seems to produce the same effect in the Narmada valley even as no national boundaries are being crossed.

The aim of this chapter is not to legitimise displacement due to development by pointing out the possibility of constructing better compensation policies. As Swainson and McGregor (2008: 165) point out:

> [T]he idea that compensation can restore the living standards of the displaced has been employed, somewhat ironically, to legitimise a displacement-inducing development path that necessitates compensation programmes in the first place. That displacement causes hardship is hardly news to compensation researchers but the case study does show how a body of work designed to protect vulnerable

populations can be co-opted in particular contexts to legitimise the very processes and projects that make them vulnerable.

I support the Andolan's position that the first priority should be to avoid displacement, especially where marginalised social groups are the main targets of such displacement.

However, to the extent that development requires certain forms of justifiable displacements, compensation policies should be viewed not as irritants on the pathway to development, but as a key part of the formulation of effective and enabling development. Such alternative formulations are already being vigorously discussed by social movements, and instead of silencing them, governments would benefit from incorporating their insights. From a pragmatic position, Nakayama et al. (1999: 455) and Chattopadhyay (2009: 13) support the involvement of third parties in discussions of resettlement. The rebuilding of place to incorporate the displaced should be a collective project and not a burden borne by those who have been subject to losses in the name of development. There is a need to continue to focus on the consequences of the Narmada dams because the struggle against dams continues and the long-term consequences of displacement are likely to be more clear over time. There is also a need for further research on how various kinds of compensations translate into specific experiences of resettlement in order to reflect on both the possibility of diversifying strategies of compensation and the impossibility of an easy replacement of lost homes through existing forms of compensation.

References

Banerjee, A. and Raju, S. 2009. Gendered mobility: Women migrants and work in urban India. *Economic and Political Weekly*, 45(28), 115–23.

Basu, P. and Silliman, J. 2000. Green and red, not saffron: Gender and the politics of resistance in the Narmada valley, in *Hinduism and Ecology: The Intersection of Earth, Sky, and Water*, edited by C. Chapple and M. Tucker. Cambridge, MA: Harvard University Press, 423–50.

Basu, P. 2010. Scale, place, and social movements: Strategies of resistance along India's Narmada river. *Revista NERA,* 13(16), 96-113.

Baviskar, A. 1997. Displacement and the Bhilala tribals of the Narmada valley, in *The Dam and the Nation: Displacement and Resettlement in the Narmada Valley*, edited by J. Dreze, M. Samson, and S. Singh. New Delhi: Oxford University Press, 103–35.

Center for Social Studies. 1997. Resettlement and rehabilitation in Gujarat, in *The Dam and the Nation: Displacement and Resettlement in the Narmada Valley*, edited by J. Dreze, M. Samson, and S. Singh. New Delhi: Oxford University Press, 215–35.

Chattopadhyay, S. 2009. Narrating everyday spaces of the resettled adivasis in Sardar Sarovar. *Population, Space and Place*, 16(2), 85-101.

D'Souza, R. 2006. *Drowned and Dammed: Colonial Capitalism and Flood Control in Eastern India*. Delhi: Oxford University Press.

Duflo, E. and Pande, R. 2007. Dam. *Quarterly Journal of Economics*, 122(2), 601–46.

Ford, N. 2009. ADB drives integration with hydro. *International Water Power and Dam Construction* [Online], Available at: http://www.waterpowermagazine.com/story.asp?storyCode=2053817 [accessed: 24 Sep 2011].

Franke, M. 2008. The displacement of the rights of displaced persons: An irreconciliation of human rights between place and movement. *Journal of Human Rights*, 7(3), 262–81.

Fujikura, R., Nakayama, M., and Takesada, N. 2009. Lessons from resettlement caused by large dam projects: Case studies from Japan, Indonesia and Sri Lanka. *International Journal of Water Resources Development*, 25(3), 407–18.

Gilmartin, D. 1995. Models of the hydraulic environment: Colonial irrigation, state power and community in the Indus basin, in *Nature, Culture, Imperialism: Essays on the Environmental History of South Asia*, edited by D. Arnold and R. Guha. New Delhi: Oxford University Press, 210–32.

Hakim, R. 1997. Resettlement and rehabilitation in the context of 'Vasava' culture, in *The Dam and the Nation: Displacement and Resettlement in the Narmada Valley*, edited by J. Dreze, M. Samson, and S. Singh. New Delhi: Oxford University Press, 136–67.

Harvey, D. 2003. *The New Imperialism*. Oxford: Oxford University Press.

Kedia, S. 2004. Changing food production strategies among Garhwali resettlers in the Himalayas. *Ecology of Food and Nutrition*, 43(6), 421–42.

Klingensmith, D. 2003. Building India's 'modern temples': Indians and Americans in the Damodar Valley Corporation, 1945–60, in *Regional Modernities: The Cultural Politics of Development in India*, edited by K. Sivaramakrishnan and A. Agrawal. Stanford, CA: Stanford University Press, 122–42.

Lawson, V. 2000. Arguments within geographies of movement: The theoretical potential of migrants' stories. *Progress in Human Geography*, 24(2), 173–89.

Lubkemann, S. 2008. Involuntary immobility: On a theoretical invisibility in forced migration studies. *Journal of Refugee Studies*, 21(4), 454–75.

Maitra, S. 2009. Development induced displacement: Issues of compensation and resettlement – Experiences from the Narmada valley and Sardar Sarovar Project. *Japanese Journal of Political Science*, 10(2), 191–211.

Multiple Action Research Group (MARG). 1991. *Sardar Sarovar Oustees in Madhya Pradesh: What Do They Know?* Vols I–IV. Delhi: MARG.

Mehta, L. and Srinivasan, B. 2000. *Balancing Pains and Gains. A Perspective Paper on Gender and Large Dams*. Capetown, South Africa: World Commission on Dams.

Moore, J. 2008. Ecological crises and the agrarian question in world-historical perspective, *Monthly Review,* 60(6), 54-63.

Nakayama, M., Gunawan, B., Yoshida, T., and Asaeda, T. 1999. Resettlement issues of Cirata dam project: A post-project review. *Water Resources Development*, 15(4), 443–58.

Nielsen, A. 2008. Political economy, Social movements and state power: A Marxian perspective on two decades of resistance to the Narmada dam projects. *Journal of Historical Sociology*, 21(2/3), 303–30.

Parasuraman, S. 1997. The anti-dam movement and rehabilitation policy, in *The Dam and the Nation: Displacement and Resettlement in the Narmada Valley*, edited by J. Dreze, M. Samson, and S. Singh. New Delhi: Oxford University Press, 26–65.

Parasuraman, S., Upadhyaya, H., and Balasubramanian, G. 2009. Sardar Sarovar Project: The war of attrition. *Economic and Political Weekly*, 45(5), 39–48.

Patwardhan, A. and Dhuru, S. 1995. *A Narmada Diary*. New York: Icarus Films.

Raghuram, P. 2009. Which migration, what development? Unsettling the edifice of migration and development. *Population, Space and Place*, 15(2), 103–17.

Ramachandra, K. 2006. Sardar Sarovar: An experience retained? *Harvard Human Rights Journal*, 19, 275–81.

Read, P. 1996. *Returning to Nothing: The Meaning of Lost Places*. Cambridge: Cambridge University Press.

Roy, A. 1999. *The Greater Common Good*. Mumbai, India: India Book Distributor.

Skeldon, R. 2006. Interlinkages between internal and international migration and development in the Asian region. *Population, Space and Place*, 12(1), 15 30.

Sunil and Smitha. 1996. Fishing in the Tawa reservoir: Adivasis struggle for livelihood. *Economic and Political Weekly,* 31(14), 870-2.

Swainson, L. and McGregor, A. 2008. Compensating for development: Orang Asli experiences of Malaysia's Sungai Selangor dam. *Asia Pacific Viewpoint*, 49(2), 155–67.

Tata Institute of Social Sciences (TISS). 1997. Experiences with resettlement and rehabilitation in Maharashtra, in *The Dam and the Nation: Displacement and Resettlement in the Narmada Valley*, edited by J. Dreze, M. Samson, and S. Singh. New Delhi: Oxford University Press, 184–213.

Thukral, E. 1996. Development, displacement and rehabilitation: Locating gender. *Economic and Political Weekly*, 41(24), 1500–03.

United Nations. 2009. *Human Development Report: Overcoming Barriers, Human Mobility and Development*. Houndmills, Hampshire, UK: Palgrave Macmillan.

Vikas. 2001. State and people's initiatives: Experience of Tawa Matsya Sangh. *Economic and Political Weekly,* 36(49), 4527-30.

Whitehead, J. 2000. Monitoring of Sardar Sarovar resettlees: A further critique. *Economic and Political Weekly*, 35(45), 3969–76.

Worster, D. 1985. *Rivers of Empire: Water, Aridity, and the Growth of the American West*. New York: Pantheon.

Xaxa, V. 1999. Transformation of tribes in India: Terms of discourse. *Economic and Political Weekly*, 34(24), 1519–24.

Yeoh, B., Willis, K. and Fakhri, S. 2003. Transnationalism and its edges. *Ethnic and Racial Studies*, 26(2): 207–17

Chapter 6
Globalization and Occupational Displacement: Indian Artisans in the Global Economy

Timothy J. Scrase

Introduction

In much of the displacement literature, the focus is on describing and analysing the loss of land and access to resources, and the subsequent social and political struggles for those displaced to sustain a livelihood. While numerous studies of displacement provide incontrovertible evidence as to the nefarious effects on impoverished urban and rural populations of over-industrialisation and over-consumption, widespread urbanisation, mass deforestation and pollution, and growing income inequalities, there have been relatively few studies that have outlined the multifarious and interlinked processes involved in occupational displacement. In other words, while the loss of one's land or access to one's traditional hunting and fishing grounds are devastating for various rural communities, these forms of loss can only partly explain the displacement facing various other communities that are located across both urban and rural sectors, and whose lives and occupations have now become deeply embedded in the vagaries of national and international economies.

The aim of this chapter is to outline and explain the complexity of what is termed occupational displacement—a form of displacement which may in fact involve a range of determining factors that have, initially, slowly evolved but which have now intensified and become more complex due to both globalization and the various intended and unintended consequences of rapid social change as a result of globalization. In analysing occupational displacement, the case of Indian artisans will be presented as an exemplar highlighting the intricacies and intertwined determinants which have led to the fragility of social life of numerous artisanal communities throughout the global south. The value, traditions and skills of Indian craftsmen and women are renowned both in India and abroad:

> Handwoven Indian textiles appear on the ramps in Paris, handcrafted Indian jewelry is sold in the best stores in New York, and handmade Indian carpets cover some of the most elegant floors in the world. The craftsmen and craftswomen who create them often have learned their art as a hereditary profession and are taught from infancy. Some skills are so intricate and so specialized (such as the

> famous *thewa* gold filigree-on-glass jewelry or the grinding of local stones and minerals into paint pigments) that the manufacturing process is a secret still closely guarded by a small number of families. Others acquire their individual luster through lifelong apprenticeship and practice. Some are regional specialties, whose techniques, motifs, and materials make them instantly identifiable; others are found, with some variation, in communities throughout India. What all the many thousands of beautiful and unique craft expressions in India have in common, though, is that the weavers, potters, carvers, painters, embroiderers, goldsmiths, and others who create such beauty with consummate skill and knowledge enjoy few of the fruits of their labor (Liebl and Roy 2004: 55).

The imbalance between the skills and knowledge of the creators—the artisans—and the dismal return on their craft skills in terms of a fair wage and adequate livelihood are indeed worrying and necessitate more thorough explication. In so doing, the chief aim of this chapter therefore is to present a model of artisanal displacement and highlight its various features by way concrete and recent examples from diverse Indian artisan communities. For the sake of clarity, crafts will refer both to handicrafts (including jewellery, pottery, wood and metal objects, and various trinkets) as well as handloom (hand woven cotton and silk as well as embroidery).

Artisanal Displacement: An Outline

Third world artisans, it has been argued, engage in a form of precarious production defined by loss of land, resources and markets as well as various health concerns with imminent financial fragility or ruin for many (see Scrase 2003). Moreover, along with their occupational displacement there is the loss of their craft skills, the craft artefact disappears, and ultimately there is cultural decline. In sum, one can identify at least ten inter-related reasons for artisanal displacement as indicated in Figure 6.1. These are: globalized production; rising costs; prices of raw materials; loss of land; loss of markets; loss of craft culture; selective traditionalism; design copying; global competition; and mass production.

The following sections of this chapter explore these interrelated aspects of artisanal displacement. In certain circumstances, such displacement, of weavers for instance, has its origins as far back as the time of 18th century British industrialisation with the mass production of cotton cloth in the mechanised mills of Northern England. Yet, the examples presented hereafter serve to highlight the intersection of fundamentally globalisation-induced problems particularly in terms of under-utilised, redundant and displaced labour, irregular or infrequent demand for products, lack of adequate protection for artisans and their crafts, and increased competition due to the opening-up of local markets under neoliberal trade policy. In other words, the widespread displacement of artisans from their traditional occupations is rapid and seemingly irreversible.

Figure 6.1 Modes of Artisanal Displacement

Globalized Production, Mass Production and Rising Costs

The globalization of the production of crafts raises a myriad of concerns and problems for individual artisans and local craft communities. Fundamentally, the reorganization of global trade arrangements has led to the opening-up of various markets, including those for handicrafts, jewellery and woven cloth. As a consequence, there is now in Asia intense competition on behalf of both large and small scale artisanal producers for gaining and retaining markets for their goods. Such competition has been reported in the case of India, for example, where craft workers complained of increasing demands placed on them for new designs, better quality product, and more reliable supply (see Scrase 2003).

The world of the marginal artisan has thus now become a world of competitive entrepreneurialism whereby there are increasing pressures to compete with each other, as well as against overseas importers, for market share and ultimately for their survival. Complicating this scenario is the fact that the national and global markets for most crafts are unpredictable and fashion-driven, dominated by the buyers from large department stores in the global North or else powerful local middlemen at the point of production. In this context, in a study of grass basket makers in Bangladesh, it was shown that: 'The variety of middlemen involved in getting the baskets from the port in Chittagong to the customer in America or Europe absorb from 80% to 90% of the final price (Poe and Kyle 2006: 16). In a different context, but also illustrating the power of the trade middleman who control the market, Esperanza's (2008) ethnographic study of Balinese artisans contracted to produce native American handicrafts ('dreamcatchers'), selected Moroccan furniture, and African drums is significant. In the process of 'outsourcing otherness', as she terms it, Esperanza demonstrates the nuanced way powerful middlemen play by controlling the markets of demand and supply for these products.

Mass production, cheap imports and virtual marketplaces are now emblematic of the globalized production of crafts. These are having a profound effect not only on the lives of individual artisans, but also on the culture of the consumers themselves. In terms of the impact of mass production and cheap imports, consider the following example, reported on the front page of the Indian newspaper, *The Telegraph* (14 August 2005) and headed, rather satirically: 'Gods, now made in China – idols from communist neighbour flood India':

> 'Containers are landing in Mumbai by the dozens every month. Not a single idol goes unsold; there's a mad scramble for them. I'm struggling to cope with the demand,' said Balwant Singh, who runs a gift shop in Mohali.
>
> 'The buyers come and ask for images of different gods and goddesses, but will accept only those made in China. Not many buy Indian-made idols now.'
>
> What makes the Chinese idols so attractive?
>
> 'Their finish is excellent. They are made of synthetic material and are very colourful,' said another gift shop owner in Chandigarh, Inder Kumar Sethi. 'The customer would take one look at a Chinese idol and immediately settle for it.'
>
> 'There is also more variety in these idols,' Singh continued. 'They are unbreakable and can be washed. The Indian ones are heavier and not as well polished. Their shelf-life is very short but the price is cheap.'
>
> The ones paying a heavy price are the manufacturers in Moradabad, Meerut, Hyderabad and Jaipur.
>
> 'They (the Chinese) have taken over our market for toys, cutlery, nail-cutters and many other items. Now they have taken over our gods as well,' a dealer rued. 'We should ban their import.'
>
> From Chandigarh, the Ganeshas, Lakshmis and Kalis [various Gods] have been finding their way to religious centres like Katra in Jammu and Kashmir, Amritsar in Punjab and Kullu in Himachal Pradesh, where the demand is high (Singh 2005).

The ever present threat of an enveloping East Asian economy, geared to rapid mass production, expediency and low cost looms large in the mindset of not just the small scale producers and entrepreneurs, but is identified as a serious economic challenge to be dealt with at the government planning level in India. With respect to crafts, the Government of India, Planning Commission (2009: 132) has observed that:

> Modern handicrafts are essentially of a mass scale, machine-made, labour-intensive items, which find their way to countries offering lower wages. Hong Kong and Taiwan rated China as the best source for an offshore production base. Items which are of higher quality and sold to upmarket consumers are still produced in Taiwan and Korea. Where design is important, the items are produced locally. Importers/customers of Indian handicrafts often do not like the items, which finally hit the export market; they may not be the 'best' of Indian handicrafts. India's overall image in foreign markets has generally been based on high volume and low value products.

The image of the 'cheap' Indian handicraft, with fading colours, poor stitching, irregular design, and a generally tattered image, works against the interests of the Indian artisans who, if working in a fair market situation, aim to perfect their work and emphasise their skills and abilities in the craft itself. But under pressure to meet impending and unrealistic market demands for international orders on time and free of imperfections, then the rush to market their wares works against their interests and inevitably orders are lost, payments are not made, and occupations, indeed certain crafts, disappear.

Prices of Raw Materials and the Loss of Land

The growing scarcity of raw materials such as clay and wood, and the high costs of various precious metals and gems, is having a profound impact on artisan lives. The loss of land though rapid urbanization, debt or forced eviction, means a loss of access to natural resources which may be essential to sustaining many artisan crafts. In relation to brass metal work, Talukdar (2007: np) reports on the case of artisans in Assam who face the combined problems of increasing prices for the raw materials, loss of income, and dwindling consumer demand:

> This long tradition now faces great economic uncertainty. Somim [a brass metal artisan] say his monthly income has dropped from Rs.3000 a couple of years back to Rs.1500 now, sometimes even less. The condition of other artisan families of Hajo is no better. Each of these families is entirely dependent on earnings from brassware manufactured in individual production units, and they've all been struggling. Somim knows that the days without job work are worsening his plight. Even to procure the allotted 33 kgs of rice from the fair price shop is becoming difficult, let alone the treatment of his ailing mother,

> or other liabilities like arranging marriages for his two sisters. His father died several years ago due to acute stomach ailment and practically without any treatment. 'The price of each kilogram of brass sheet has gone up from Rs.150 in April 2005 to Rs.320 in 2007, posing a serious threat to the survival of the artisan families and their unique tradition of brass metal craft, which has a deep root in Assamese culture and heritage', says Md. Saiful Majid, manager, Hajo Brass Utensils Workers Co-operative Industries Society Ltd, one of the oldest co-operatives in the area, founded in 1952. Meanwhile, the daily remuneration to an artisan has remained frozen at Rs.100 since 2005, and only 10-15 days of work are available in a month, against an average of 25-27 days earlier. Even from this fixed remuneration of Rs.100 the artisans have to bear other related expenses incurred in procuring waste coal from the steel factories and chemicals needed for manufacturing the products. Somim does not blame the *mahajans* [middle-men] alone. 'When there are frequent hikes in price of the raw materials, it is difficult for them also to afford money. Also, with raw material prices higher, the cost of making things is higher, but buyers are not willing to pay more. This has led to a gradual decline in our monthly income', he says.

With respect to handloom, over the past decade or so price rises of various kinds, together with a general lack of government support or adequate subsidy to the handloom sector, has had devastating consequences for a craft that conservatively would support upward to 12,000,000 households directly and four times that number indirectly. Reddy (2006: np) explains the complexity of the price increases that occur at several levels:

> [...] yarn prices are steadily increasing. The availability of hank yarn —the basic material from which weaving is done —is a serious issue because it is controlled by modern spinning mills, who see more profit in large-volume cone yarn. Secondly, since hank yarn is tax-free and has subsidies, enormous amounts are diverted to the powerloom and mill sectors. As a result, there is a perennial shortage of yarn for the weavers. Despite a few schemes, the hank yarn access issue has not been resolved. Colours are expensive, and presently there is no system or mechanism to increase their availability. [Moreover] handloom primarily uses natural fibres such as cotton, silk and jute. Prices of these fibres have been increasing during production and processing. Cotton production in India is expensive because of intensive and high usage of costly agricultural inputs such as pesticides and fertilisers. Secondly, while the fibre production most often happens in the vicinity of the weavers, their processing is done in distant areas, and as such the prices to the weaver are higher. With the central government now encouraging primary fibre and yarn exports, handloom weavers would be on the last priority for yarn suppliers.

Under this rather dismal scenario it is little wonder then that artisans and their offspring do not picture a viable future in craft work – one involving arduous training, long days, dangerous and often dirty work and for little financial reward.

Global Competition and Design Copying

The increasing global competitiveness for market share of crafts at ever decreasing prices continues unabated. According to Fowler (2004: 113), 'Global competition to provide products at the lowest possible price point has proliferated to counterfeiting of original handmade crafts. An example of this is conveyed through the sale of Native American arts and crafts that are actually imported copies from Asia'. Problems of design copying flourishes in the Indian handicrafts sector, as they do throughout the global trade in crafts, even to the point of copies of copies being produced and sold as original. According to Page (1998, cited in Fowler 2004: 113):

> Some [crafts] are of sufficient quality to be virtually indistinguishable from the originals by all but the most practiced eye. But they *are* almost always distinguishable by one factor: price. Southwestern Indian-type basketry is now made in Pakistan, and Romania has begun manufacturing and selling knockoffs of Taiwanese knockoffs of Indian jewelry. In almost every case, the prices of such items are less than what would be charged for authentic material (original emphasis).

The Planning Commission of the Government of India was well aware of increasing global competitiveness in the handicraft sector when it wrote in its 2009 report that:

> India's closest competitor in the handicrafts sector is China. India-China comparative strengths in some of the important sectors are:
>
> **Embroidery:** India leads in USA and UK, China dominates elsewhere.
> **Stone and Wood Products:** China overwhelms India everywhere.
> **Hand-knotted Carpets:** India is a leading supplier to Germany whereas it trails much behind China in exports to the US.
> **Woven and Other Carpets:** India is the main source.
> **Metalware:** India has a smaller, though significant, share in USA and Germany.
>
> China's carpet industry has been concentrated in large factories. This has enabled China to have higher degree of mechanisation and better control over delivery schedule (2009: 133).

Both technological advances in neighbouring countries such as Taiwan, South Korea and Hong Kong, along with geographical advantage, give China the edge in dominating the world production and trade in crafts, as well as cloth and ready-made garments. As the Planning Commission report commented, '... for imitation jewellery, China imported technology from Taiwan and Korea. For brassware, Korea exported technology to China, e.g., for manufacture of picture frames. For toys and dolls, Korea and Taiwan provided technology to China and Thailand' (2009: 132).

The extent of the impact of mass production and blatant copying on artisan lives is found in the case of silk sari weavers of Benares. Emily Wax, in her interviews with weaving families, relates the following:

> This sari design, which has been in Javen's family for 100 years, can take up to two months to weave. Patterns like these have been a source of Indian pride for more than 2,000 years, with India's version of haute couture adorning wealthy women of the empires of Rome, Egypt and Persia. Until recently, weaving was India's second-most-common occupation, behind farming. But in this ancient city along the Ganges, Hinduism's holiest river, an estimated 1 million sari weavers are facing almost certain ruin. Cheaper, machine-made saris – many of which are copied from Varanasi's famous patterns – are being pumped out of China and from newer factories in India's western Gujarat state. Adding to the weavers' woes, changing fashions and global trade rules have opened the Indian market to foreign competitors, leaving many once-prosperous sari weavers and their families in desperate poverty. 'This loom will be in a museum', said Javen's despairing uncle, Nazir Ahmed, 30, whose family was forced to shut down 12 of their 14 looms. 'We would have never predicted this. We were India's artists. Now we are living in poverty.' (Wax 2007:np).

A similar story concerning the loss of skills, markets and, for many, even their lives, was reported a year later on the BBC website in an article titled: 'Can Indian embroidery resist Chinese threat?' (Pandey 2008). Rather than Benares, the BBC reporter Geeta Pandey details the downward spiral of the Lucknow *chikan* or *chikankari* (intricate thread embroidery) industry in the face of cheap overseas imports which, to many consumers it seems, are of equal quality to the local handmade variety. As explained in the report: 'A shopper takes keen interest in an orange-green shirt. 'It's very pretty. I really like it and would love to buy it for my daughter. But this size is too big. Shame they don't have her size', she says. It's obvious the made-in-China tag doesn't seem to bother the customer (Pandey 2008:np).

One wonders whether the Indian consumer also realises the enormous transformation of the *chikan* industry that has occurred over the decades. Indeed, 'In each generation, the struggle to salvage *chikan* from its condition of decline is invented anew, with previous efforts apparently forgotten' (Wilkinson-Weber 2004:290). Unlike earlier times, in a patriarchal turn, *chikan* has become a declining craft now dominated by women at the producer level, but is now organised and controlled by men, many of whom were formerly the embroiderers. These nuanced transformations in the rise and fall of various Indian crafts remain to be fully researched, yet they highlight the twin perils for artisans – global competition, copying and cheating on the one hand; the inevitable disappearance of one's craft occupation due to 'natural attrition' and decline on the other.

Research I conducted in early 2003 with artisans in New Delhi also exposed their frustration with design copying not only from abroad, but even by one's 'fellow' artisans. While there is a recognition for modification of craft styles to suit changing market demands, there is a certain inevitable consequence of design copy, with eventual market saturation. As one artisan expressed it, after attending a workshop on 'Design Development':

> [...] everything spreads in the market. An instructor from NID (National Institute of Design) gives a series of training workshops in five places and the same design gets popularized in the market in the same period of time, thus becoming too common and of no value. There is no profit in such training. Instead, our traditional designs are much better – the horse, Mother Goddess – these are uncommon to the terracotta craftsmen from other states - Traditional and our own designs are the best for earning profit from the market.

These comments certainly point to the need for originality in design and a retention of one's design integrity, but also for better practice and awareness on behalf of the NGOs and government training bodies.

Loss of Markets, the Loss of Craft Culture, and Selective Traditionalism

One could argue that globalisation and competitive trade, together with the internet and e-commerce, has increased substantially the markets for Indian crafts. Yet, as has been argued above, and emphasised in a range of examples, working out precisely which crafts and which artisan communities have benefited over the past two decades or so of neoliberal free trade arrangements remains contentious, complex and difficult to estimate. Mass production and copying of designs has certainly impacted heavily in the handloom sector. Moreover, craft sustainability relies extensively on allied or partner industries such as tourism (particularly handicrafts, pottery and cloth) and fashion (jewellery, embroidery and weaving). When demand is high and orders are to be met at short notice, then the time taken to produce the craft is reduced hence various production techniques (machines, sub-contracting, importing) are employed at the production level to ensure the craft is in the market or in the hands of the buyer as expeditiously as possible. In that way, through various production shortcuts, craft expertise – the embodiment of the artisan in the craft – is devalued, even though the item itself may look similar or even better than the original, fully hand-made piece.

The intricacies and organisation of artisan lifeworlds is also of significance. While various craft cultures are in decline or re-inventing themselves, understanding the 'doing of craft' is especially important. This is emphasised in a study of Pattamadi mat weavers of Tamil Nadu who resort to various alliances and kinship arrangements to maintain a steady livelihood. As Venkatesan (2006: 76) explains:

> Weavers constantly judge their work against that of others. Weaving is embodied work – weavers pride themselves on the skill in their hands and the keenness of their eyes (*kann nidhanam*) in matching colours and ensuring symmetry in the patterning of the mats.

Thus, rather than being a somewhat cooperative and harmonious community, Venkatesan's study exposes the nuanced divisions and patterns of patronage and alliances that determine the daily fortunes of many artisan families. He goes on to write that:

> The private traders compete with each other for mats and for the loyalty of weavers. Weavers who form part of a trader's household, his wife and children, usually supply mats to him. Others require a mixture of patronage and coercion to continue supplying mats to the same trader. As the bulk of weaving takes place in homes and weavers often own the raw materials they require, they are theoretically free to sell their mats to any buyer. In practice traders constantly seek to limit this freedom. This is done in different ways. For instance, a trader might give a loan to a weaver in times of need, which is then recovered in the form of mats over a period of time. Traders also try and avoid paying a weaver the total sum of money owed for a mat (Venkatesan 2006: 81).

The resilience of Indian crafts, despite global forces rendering various crafts redundant, is noteworthy. One interpretation of this fact is that the resurrection and increasing popularity of 'native', 'traditional' or 'authentic' Indian arts and crafts can be seen in the context of the Indian urban elite and diasporic community of professional migrants seeking to interrogate and reconstruct their identities, through literature, film and art, in a fragmented and globalizing world (Appadurai 1996). Indeed, certain 'traditional' crafts retain their popularity and cultural value, and benefit from state patronage and support by their inclusion in various local, national and international fairs and festivals as well as displayed in museums or on sale in state emporia.

In this context, a schism has evolved between elite and quotidian crafts in India. Jain (1995), for instance, argues that a great divide now exists between the mainstream elite artists who work in studios and sell through organized galleries and the 'everyday' artisans and craftspeople. This latter group:

> [...] mainly thrive on the new urban patronage which has arisen as a result of protection and patronizing developmental endeavours on the part of the government. The government encourages these with a view to keeping the

> artisans self-employed and to earning precious foreign exchange by exporting manufactured craft products. Once it is established that the 'crafts' are primarily 'commercial' rather than 'cultural' in nature, their treatment involves different strategies, one of which must be that the criteria for design and aesthetics are oriented to commerce-related development (Jain 1995: 29).

The emergence of what can be termed selective traditionalism (of crafts) is seen in the context of an historical, romanticized view of the traditional Indian life, one rooted in the past in contrast to the 'advanced' world of industrial Britain, the colonizer and bearer of modernity. In post-independent India, such views were reinforced by state patronage of various crafts and their display in, for example, the Indian Crafts Museum where, it has been observed by Greenough (1995) that continuity and survival within India's material tradition, rather than innovation, determines whether an artefact is accepted as an addition to the collection.

The failure to recognize 'on-the-ground', local knowledge and incorporate this into employment policies and planning shows a disdain for marginal workers and so reproduces the failures of top-down policy making so indicative of ill-conceived, developmentalist policies. Moreover, questions concerning 'who decides what activities are worth supporting?' and 'what specific crafts are considered unique to a nation and its peoples and why' are significant . In this sense, the state involves itself in 'selective traditionalising', a process which can assist some communities to survive but may condemn the fate of many others.

At the more mundane level, Indian artisans, when interviewed in 2003, were often highly critical of government support for several reasons. Petty corruption was at the forefront of their complaints as the following account reveals:

> *Interviewer*: What other support is there from the government?
> *Artisan response*: To get to a fair we have to give money then only do we get an approval letter. Again we have to pay money to the officers at the fair.
> *Interviewer*: How much?
> *Artisan response*: For a letter it is 1500-5000 rupees.
> *Interviewer*: Receiving the letter confirms your participation in the fair?
> *Artisan response*: No. To get information about the fair we have to pay more money. We are ready [willing] to pay to get to a fair. Government organizes fairs free of cost yet we still have to pay the officers. We are also entitled for TA/DA [travel and dearness allowances] but that money goes to officers' pockets in addition to the other money we pay. If someone else gives more money than us then the [approval] letter will be issued for him and not for us. So even if we pay bribes, our participation in a fair is not confirmed.

Corruption is not just confined to participation in fairs but is virtually in all stages and levels of the display and marketing of crafts.

> *Interviewer*: Do the wholesalers send crafts to *Manjusha* or *Tantuja* [various state government emporiums]?
> *Artisan response*: If you know any officer in the cottage emporium then you can enter your crafts by bribing them. They are all corrupted. You also have to bribe the 4th class staff to display your crafts so that a sale can possibly take place, especially as the manager is not going to help you. Payment is on the sale basis and you earn a check every month. If no craft item is sold then they will return it to you after 3-4 years.
> *Interviewer*: Are societies [NGOs] more helpful in getting information and reciprocation for fairs?
> *Artisan response*: Artisan societies are there. But now all the channels are corrupted. To make a society 10-15 members are needed and are fulfilled by own family members and friends. Whatever money is sanctioned by the NGO or other public/private organizations is then put into personal pockets.

Thus, for many artisans in India, problems of corruption by NGOs and government officials, together with a bias to certain craft styles or artisanal groups, are endured daily in their efforts to raise the profile of their crafts as well as to simply earn a living.

Conclusion

The foregoing discussion has presented a threatening picture for Indian artisans whose livelihoods are under constant risk from largely globalised-induced trade inequities, and a fickle and demanding consumer market. In the model of artisanal displacement presented earlier, the interrelationship between its 10 features has been highlighted, and examples drawn upon, to emphasise the fragility and precarious nature of craft workers, their families and their uncertain futures. And yet, in spite of such forces operating against artisans, there are many examples of innovation and a resurgence of dying craft skills. In the state of Andra Pradesh, the ebbs and flows of woodcraft is explained by Deshpande (2011: np):

> By the mid-1980s, employment opportunities in the local sugar mills lured craftsmen away with the promise of higher incomes. … Chitti Raju encouraged the few remaining practising craftsmen by introducing them to exhibitions that were being held across India to promote traditional crafts. Through these exhibitions, a new world of opportunities became available to the artisans of Etikoppaka, where they could imbibe fresh ideas in design. Despite these efforts, woodwork artisans were living on the edge in a perpetual battle against poverty, with migration as the only recourse left … Raju recognised that the legacy of woodcraft in Etikoppaka could soon be relegated to the confines of memories. To bolster the artisans and their skills, he directed his efforts towards obtaining a fair remuneration for their work in both national and international markets. He diversified the product range by introducing toys and decorative pieces, and focussed on improving the quality of the artefacts. Eventually, his efforts bore

> fruit: the uprooting of artisans from their soil and skill was stemmed. However, when a few importing countries rejected their products in the early nineties because of the lead content in their synthetic dyes, Raju had a new problem on hand. Determined to find a way out of the predicament, Raju devoted himself to reviving the age-old practice of using dyes extracted from plants and trees. He explored the fascinating world of trees and the secret dyes in their roots, bark, leaves, rhizomes, fruit and seeds. Beginning with procuring natural dyes from sources outside the village, he was soon on his way to discovering alternative local sources as well. Today he animatedly counts fifteen natural dyes in use in Etikoppaka, including the majestic red of Indian madder and the deep blue of Indigo. These lead-free natural dyes gave a new lease of life to craft in the village.

The switch to environmentally friendly and safe techniques was both fortunate and timely, with a growing 'ethical' consumer demand and interest in alternative and green consumption and fair trade. While such 'ethically-produced' handicrafts are relatively few in number, they are finding their way into the mainstream local and global marketplace via the internet and various NGO shops that stock and sell their products. This has not been without criticism, in the sense that often there is ethnic or racist stereotyping involved in the marketing of these goods, as well as a degree of commodification of poverty in the sometimes blatant selling of crafts at 'discount' rates or in the advertisements that accompany these products (Scrase 2010).

Nevertheless, for artisans seeking out new markets for their products, especially with the assistance of ethical NGOs and operating in the world of fair trade and with realistic production and marketing strategies, then their future is rather more viable, as the research by Littrel and Dickson (2010) reveals. It is through a combination of craft innovation, appropriate government policy and financial support, adequate training, and cooperative development that the problems of occupational displacement can be somewhat ameliorated.

References

Appadurai, A. 1996. *Modernity At Large: The Cultural Dimensions of Globalization*, Minneapolis: University of Minnesota Press.

Deshpande, N. 2011. Where woodcraft is a way of life. 4 December. [online]. Available at: http://www.indiatogether.org/2011/dec/eco-etik.htm [accessed: 10 December 2011].

Esperanza, J. 2008. Outsourcing Otherness: Crafting and Marketing Culture in the Global Handicrafts Market, in *Hidden Hands in the Market: Ethnographies of Fair Trade, Ethical Consumption, and Corporate Social Responsibility*, edited by G. deNeve, et. al. Bingley: Emerald JAI Press, 71-96.

Fowler, B. 2004. Preventing Counterfeit Craft Designs, in *Poor People's Knowledge: Promoting Intellectual Property in Developing Countries*, edited by J.M. Finger and P. Schuler. Washington: The World Bank, 113–131.

Government of India, Planning Commission. 2009. *Uttar Pradesh: State Development Report (Volume II)*, [online]. Available at: http://planningcommission.nic.in/plans/stateplan/index.php?state=sdr_up.htm [accessed: 2 November 2011].

Greenough, P. 1995. Nation, economy and tradition displayed: the Indian Crafts Museum, New Delhi, in *Consuming Modernity: Public Culture in a South Asian World*, edited by C.A. Breckenridge. Minneapolis: University of Minnesota Press, 216-248.

Jain, J. 1995. Art and artisans: Tribal and folk art in India', in *The Necessity of Craft*, edited by L. Kaino. Nedlands: University of Western Australia Press, 24-34.

Liebl, M & Roy, T 2004, 'Handmade in India: traditional craft skills in a changing world', in *Poor People's Knowledge: Promoting Intellectual Property in Developing Countries*, edited by J.M. Finger and P. Schuler. Washington: The World Bank, 53–73.

Littrell, M.A. and Dickson, M.A. 2010. *Artisans and Fair Trade: Crafting Development*, Sterling VA: Kumarian Press.

Pandey, G. 2008. Can Indian embroidery resist Chinese threat? *BBC*, 3 March. [online]. Available at: http://news.bbc.co.uk/2/hi/south_asia/7232238.stm [accessed: 12 May 2009].

Poe, K. and Kyle, S. 2006. Fair Trade – Is It Really Better for Workers? A Case Study of Kaisa Grass Baskets in Bangladesh, Working Paper, Department of Applied Economics and Management, Cornell University, Ithaca, New York.

Reddy, N. 2006. Weaving woes on the handlooms, 7 February. [online]. Available at: http://www.indiatogether.org/2006/feb/eco-handloom.htm#continue [accessed: 17 October 2010].

Scrase, T.J. 2010. Fair Trade in Cyberspace: The Commodification of Poverty and the Marketing of Handicrafts on the Internet, in *Ethical Consumption: A critical introduction*, edited by E. Potter and T. Lewis, London: Routledge, 54-70.

Scrase, T.J. 2003. Precarious production: globalisation and artisan labour in the third world. *Third World Quarterly,* 24(3), 449-461.

Singh, G. 2005. Gods, now made in China —Idols from communist neighbour flood India, *The Telegraph* (Calcutta), 14 August. [online]. Available at: http://www.telegraphindia.com/1050814/asp/frontpage/story_5113622.asp [accessed: 21 August 2010].

Talukdar, R.B. 2007. Brass metal work losing its shine, India Together, 21 November. [online]. Available at: http://www.indiatogether.org/2007/nov/eco-brass.htm [accessed: 9 July 2010].

Venkatesan, S. 2006. Shifting balances in a craft community: The mat weavers of Pattamadai, South India. *Contributions to Indian Sociology* (n.s.) 40(1),63-89.

Wax, E. 2007. An ancient Indian craft left in tatters: sari weavers struggle amid economic boom, *Washington Post Foreign Service*, 6 June [online]. Available at: http://www.washingtonpost.com/wpyn/content/article/2007/06/05/AR2007060502858.html [accessed: 29 April 2008].

Wilkinson-Weber, C. 2004. Women, work and the imagination of craft in South Asia. *Contemporary South Asia*. 13(3),287-306.

PART 2
Politics and Coercive Displacement: States Displacing Minorities and Indigenous Peoples

Chapter 7
The Land Issue on Banggi Island, Sabah, Malaysia: Deagrarianisation and Exclusion of the Bonggi

Fadzilah Majid Cooke[1]

Introduction

This chapter is concerned with long-term changes in Bonggi livelihoods in the villages where Bonggi communities live on the island of Banggi in northwest Sabah, Malaysia, which ranges from a complete departure from subsistence production in favour of cash crop production to one of spatial and sectoral diversification. This chapter benefits from debates on deagrarianisation that in general has altered views concerning the contours of poverty including its causes. According to the deagrarianisation perspective, spatial diversification (delocalisation) of livelihoods occurring through ex-situ employment (often via migration) and sectoral diversification through non-farm employment opportunities both in-situ or ex-situ can provide exit conditions that are amenable to an escape from poverty. Notably, Bryceson (2002) has written on deagrarianisation in the African continent, which takes account of the agrarian transition facing many rural areas captured in the transformation of livelihoods and income structures at the household and community levels. This transformation is such that livelihood and income structures may no longer be wholly dependent on agriculture. These dramatic

1 This paper is part of a larger and longer term project that the author has undertaken over a period of several years with Bonggi of Banggi Island, in the district of Kudat, Sabah. Work on Bonggi are included in such publications as Majid Cooke, F. and Justine Vaz (2011) *A review of indigenous peoples and community conserved areas in Sabah.* Report submitted to the Japan International Cooperation Agency (JICA) as part of the project Traditional Ecological Knowledge in Sabah under the Bornean Biodiversity and Ecosystems Conservation Phase II Programme, Global Diversity Foundation (GDF), Kota Kinabalu. The author also wishes to thank the Challenges of the Agrarian Transition in Southeast Asia Project (CHATSEA) under the Directorship of Professor Rudolphe De Konincke of the University of Montreal, Canada, for funding support to present an earlier version of this paper at the conference: *Revisiting Agrarian Transformation in Southeast Asia: Empirical, Theoretical, and Applied Perspectives*, held at Chiang Mai, 13-15 May 2010, organized by the Centre for Social Sciences and Sustainable Development, Chiang Mai.

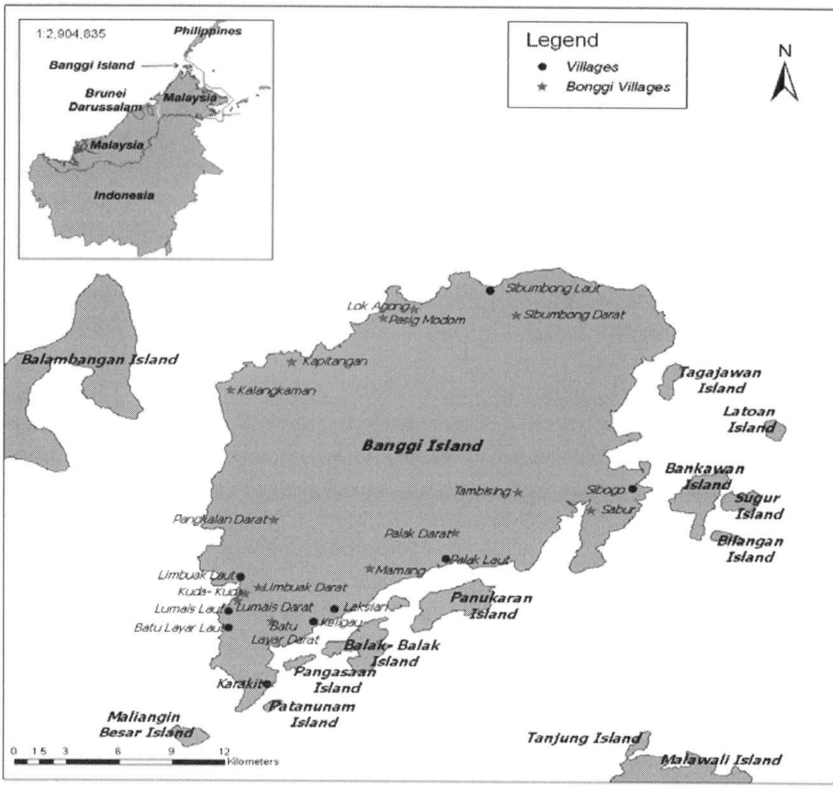

Figure 7.1 Banggi Island Highlighting Bonggi Villages

changes can occur within a generation and belie the seemingly unchanging and stagnant agricultural orientation of the landscape of many rural areas.

There is a multitude of ethnic groups living on Banggi Island. The chapter focuses on the land-based group known as Bonggi; specifically those living in five villages, namely, Batu Layar, Limbuak Darat, Kalangkaman, Kapitangan, and Sabur (see Figure 7.1).

Changes to the first four villages occurred in conjunction with a government agricultural (*Pertanian*) and resettlement scheme (referred to from hereon as the *Pertanian* scheme). The *Pertanian* scheme was aimed at converting subsistence farmers living in the scattered hills of the island and in its hinterland into small-holder agriculturists growing coconut for the production of copra. They were encouraged to leave their settlements in the hills, to be housed in planned villages and allotted 6 hectares of land per family.[2] Imagined by development practitioners

2 The allocation of 6 hectares are provided for in the Sabah Land Ordinance 1930 for lands to be awarded under Native Titles (NT), so that a single NT is awarded to approved

as being isolated and poor, the Bonggi became ripe for development strategies that were aimed at improving their lot, triggering a chain of events beginning with their losing their anchor to customary land when they joined the *Pertanian* scheme. For the fifth village, Sabur, changes occurred in conjunction with voluntary resettlement in order to make way for the private development of a cattle farm.

Livelihood change, subsequent to resettlement in the 1960s and the 1970s, was marked at Limbuak Darat and Batu Layar by diversification of the Bonggi economy through the promotion of copra production by the Department of Agriculture, combined with subsistence production of mainly tapioca, sweet potatoes, and, to a smaller extent, rice, as well as the small-scale sale of vegetables and corn. The returns from agricultural production, however, fell far short of consumption needs as the cost of transporting copra by ship to the mainland at Kudat was high and the market for the sale of agricultural produce on the island was, and remained, small.

Copra production under the *Pertanian* scheme was also embraced by Bonggi communities living in the northern villages of Kalangkaman and Kapitangan. These two villages similarly suffer from a lack of outlets, high transport costs, and, communication problems since the only road that linked the villages to the sub-district administrative centre at Kerakit (see Figure 7.1) could not be accessed during rainy or wet weather conditions.

As government subsidies for coconut production lessened because of a lack of funds in the 1980s and production began to stagnate, the overall return from agricultural production fell short of minimum consumption needs. All four villages supplemented their livelihoods through in-situ non-farm work, such as working on the island's cattle farm or in the logging industry which took place in the 1980s, and in ex-situ work via out-migration. After a succession of public-sector and state-development initiated projects, the latest type of intervention arrived in the form of a large-scale single crop rubber plantation agriculture managed by the federally linked Federal Land Consolidation and Rehabilitation Authority (FELCRA). Sabur, by virtue of its relocation, was too remote to receive state attention; it found a new livelihood, namely, the commercial production of mangrove products (poles and bark).

An earlier view equating out-migration and agrarian transition with marginalisation is critiqued by the deagrarianisation approach. Deagrarianisation views diversification of rural household economies as providing an opportunity for the poor to engage in non-farm, income-generating activities (Bryceson 2002; Rigg 2005 and 2006). Proponents point to evidence of declining returns to farming compared to the relatively higher returns obtained from many non-farm activities. In instances where incomes from non-farm work are able to provide supplementary returns to farming, farm households could chart an escape route out of poverty (Rigg 1998: 503).

The possibility of non-farm work having minimal effect and, conversely, causing an increase in poverty, is acknowledged in the literature on deagrarianisation. There

---

applications of 6 hectares, or less, more often the latter.

is a possibility that the poor could find jobs in non-agricultural employment, but such jobs might not necessarily provide them with better paying job opportunities as has happened with rickshaw pullers in Bangladesh or petty traders in Java (cited in Rigg 1998: 503–4), and in the Bonggi example, in low-paying construction and agriculture-related work. However, the overall optimism that appears to characterise this approach is more clearly expressed than the cautionary voice.

According to the deagrarianisation perspective, inequality in rural areas is no longer associated with unequal land distribution, but is a result of differential access to non-farm incomes (Rigg 1998: 513). By implication, farming households with no recourse to off-farm work are the ones deprived of a 'safety valve' because of not having access to sufficient viable options to make an exit from agriculture (Rigg 1998). Small-holder households could find new resilience through market based non-farm work if their landholdings only provided them with sub-viable options for survival (Rigg 2005: 179). The suggestions are clear, namely, that the role and importance of land has progressively declined in rural economies and that non-farm work provides the poor with a mechanism for avoiding marginalisation.

The deagrarianisation approach has policy implications. Since non-farm work can be entered into by anyone (poor or non-poor alike), the lack of access to land or the unequal distribution of land may no longer be the primary cause for poverty. It meant that off- or non-farm work bears the potential for eradicating poverty without having to address inequalities (Rigg 2005: 176).

Method

The chapter suggests that assumptions about the functionality of non-farm work needs to be verified by examining the impact of diversification on economy/ies at the household level, and over an extended period. It benefits from several years of periodic fieldwork on the island.[3]

Diversification associated with non-farm work as a result of migration to urban centres or to other rural areas, if examined at the household level and in the longitudinal term, may be a useful approach to evaluate migration as a diversification strategy and its impact at the household level. The suggestion has already been made by Rigg and Nattapoolwat (2001) in their study in northern Thailand that, in the long term, farmers who are not trapped in agricultural activities and those who have access to non-farm incomes are in fact better-off than those who remain in traditional agricultural production. With regard to the Bonggi, diversification has expanded the portfolio of jobs over three decades, but it has also increased poverty. In fact, development schemes that were meant to improve their lot is one

3 Repeated short fieldwork (two to four weeks' duration) were conducted over a period of seven years beginning 2003 as follows: 2003 and 2004 in Sabur; 2004 and 2005 in Limbuak Darat and Batu Layar; 2008 and 2009 in Limbuak Darat, Batu Layar, Kalangkaman, and Kapitangan; and 2010 in Karakit (Banggi Island's Sub-district administrative centre).

contributing factor to their marginalisation, which, as the chapter indicates, is a gradual process. One strategy for coping with marginalisation is migration.

The data on out-migration was culled using random interviews of 24 return-migrants from three villages (Kapitangan, Limbuak Darat, and Batu Layar) in November 2009. The length of time they were away from Banggi ranged from eight months to forty-two years. The respondents ranged in age from 24 to 74 years. Only four interviewees were women, reflecting the tendency for migration to be largely a male activity. The interviews were largely open ended lasting between 30 to 40 minutes with a cluster of questions dealing with: (*a*) reasons for leaving home and for returning; as well as (*b*) experiences and effects of migration on individuals and households.

By using the long-term approach and paying attention to processes at the household level, this chapter makes a contribution to the deagrarianisation literature by focusing on conditions under which poor farmers quit subsistence agriculture. Although, as noted, the deagrarianisation literature does acknowledge the possibility that diversification may also cause increased poverty under certain conditions, this aspect is glossed over as 'diversification for survival', propelled by the idea that poor people do not remain poor and can indeed get out of their condition (Rigg 1998, 2005, 2006).

The chapter outlines the deprived conditions that propelled the Bonggi exit from agriculture for non-farm work, such as a lack of formal education, political marginality, and geographical isolation not being near a vibrant urban centre as in the Rigg and Nattapoolwatt case in northern Thailand. It argues that conditions of this exit may be important in determining how well households survive in new environments.

Bonggi and Development

Banggi Island, located in the Kudat administrative district in northern Sabah, is surrounded by the Sulu Sulawesi Sea to the east and north, and the South China sea to the west. It is the largest of a chain of islands to the north of Kudat. Bonggi are a major ethnic group on the island, and they claim to be the only long-established indigenous community in the area compared to those they consider as more recent newcomers, such as Ubian and Kagayan. It is estimated that 4,000 Banggi live in 11 villages (see Figure 7.1) on an island that is home to approximately 14,000 people from six or more ethnic groups (Banggi Sub-district Administration 2004). Figure 7.1 also indicates that most settlements are situated on the coastal areas of the island. This is a relatively recent formation which has arisen from resettlement policies associated with state and private sector agricultural projects and will be discussed in detail below.

Deagrarianisation at Sabur

At every point in the history of land development on Banggi Island, Bonggi had to relocate; the agrarian transition made by Sabur is the most complete. Since the 1960s, Bonggi villages have been visited by several waves of development intervention. Having to make room for the establishment of a 4,050 hectares (10,000 acres) privately owned cattle farm, the village of Sabur lost access to lands it had customary claims to (although strictly, under common law rights remain unless they have been specifically extinguished)[4]. Supporting the development of the cattle farm, and later its expansion, Sabur leaders soon found that the only land available to them were the site of mangrove forests. The availability of this natural resource and the demand for its bark and poles at the time were sufficient to prompt a process of 'agrarian transition' beginning in the 1960s.

The literature suggests that agrarian transitions are characterised by non-farm activities becoming a major source of income for rural households (Rigg 1998: 500). The Bonggi at Sabur almost completely abandoned subsistence agriculture in favour of mangrove production. Later, when mangrove production could not supply those at Sabur with the minimum basic needs, many undertook a second exit from agriculture through outmigration. Out-migration to the nearby District capital of Kudat, and, to a lesser extent, to other parts of Sabah and Peninsular Malaysia began in the 1970s and continues until today.

In the 1980s, logging took place in order to clear the area for cattle. Sabur leadership was pro logging and cattle ranching so that some at Sabur adjusted by working in the logging camps and for the cattle ranch, while their leaders hoped to benefit from the contracts emanating from the farm.[5] However, Sabur Bonggi were not sufficiently experienced nor organised to win the contracts. From about the 1980s, their economy was transformed from one of subsistence food production (tapioca, bananas, yam, and fruit) for consumption and barter, to one of mangrove harvesting for poles (*bakau*) and bark (*tangar*) for cash. This transformation was entered into, apparently, voluntarily and it became an activity that made many even more reliant on natural resource extraction than before.

---

4 Several common law decisions in the Court systems of Malaysia over the last decade have established that customary rights to lands once established, remain, unless specific declarations or legislations are adopted to extinguish them (Bulan and Locklear 2008). Such a ruling was made by the Courts on several cases including the Nor Anak Nyawi (Sarawak) , Adong (Peninsular Malaysia) , Sadong Tasi (Peninsular Malaysia). (Bulan and Locklear 2008) Ramibilin bt. Ambit (Sabah).(Free Malaysia Today 25 March 2011 online: http://www.freemalaysiatoday.com/category/nation/2011/03/25 accessed on 9 September 2012).

5 This hope was what kept the leadership against supporting opposition political party's move from getting the land back from the cattle ranch for Sabur and for other affected villages in the 1990s (interviews held in Limbuak Darat and Sabur villages, August 2004).

What were the institutional complexities that underpin the 'choices' made by the Sabur Bonggi? I argue that 'choice' exercised by Bonggi in favour of mangrove harvesting and migration as well as economic diversification was interspersed with symbolic significance and practical considerations.

Growing coconut was not a possibility for Sabur. First, at the time coconuts were introduced to Banggi by the Pertanian schemes, Sabur Bonggi had already begun working on mangroves, although growing food crops remained an important activity at that time. The market for mangroves was local as well as international (that is, the Philippines). Locally, mangroves were cut for lumber for residential or industrial use. Many houses and bridges on Banggi use mangrove poles as support. In addition, the best of the poles from stands of bakau (*Rhizophora mucronata*) were used in urban centres as cheap piling material – cheaper than concrete ones (Majid Cooke 2003). Bakau harvesting was usually done by men, but poor widowed women also actively participated.

In addition, Bonggi also produced mangrove bark from stands of tangar (*ceriops tagal*). These were exported to the Philippines for tanning (used in floor wax) and for colouring purposes. Traders informed that the red colour from the bark was also used in the Philippines for colouring the home-made alcoholic drink *tuba* (interviews, Kudat, 2 February 2003). Bark was produced by women as well as men.

Second, Sabur was bypassed by development agents because of their geographical isolation since the only way to reach the village was (and still is) by boat, making travel difficult for agricultural extension workers from the Sub-district centre Karakit or from the government agricultural station then located at the village of Limbuak Laut. So, Sabur Bonggis' contact with administration was minimal.

Third, a clear evaluation on their part of a lack of informal institutional capacity within their own community to grow the crop contributed to their ambivalence to the production of coconut. The coconut scheme was managed through headmen or *orang tua* (or *ketua kampong* or *panglima*, as known in some places). Bonggi described their choice against coconuts or why coconuts failed to be produced in Sabur as being related to intra-community relations. 'Our leaders could not be trusted with the wages given to them for the people to work on coconuts'; or 'people refused to work because they were not united' (*Tidak bersatu padu*). This could be their way of saying that, at the time, good leadership was not available at Sabur to manage a new crop.

Fourth, the symbolic significance of mangrove production and '*makan gaji*' (waged work) was linked with ideas about 'catching up' in terms of material wealth and status. In general, Bonggi at Sabur, carry memories of exchange relations with coastal groups according to how fairly or unfairly they were treated by the 'Other', from their perspective. When talking about Ubian, stereotypes abound about their aggression and unfair trading practices (also see Boutin 1994). In the ethnic stratification that emerged from the complexities of political economy, Bonggi are ranked the lowest in the hierarchical ladder.[6] Growing crops for cash is viewed as a way of catching up and of enhancing one's status through an improved standard of living. By the year 2003/2004, apart from a few sticks of tapioca and yam that were grown in front or at the back of houses, only one family was considering agriculture seriously. The major livelihood activity has been, for some years now, mangrove harvesting.

Despite diversification via mangrove harvesting, *makan gaji* (waged work) and migration to Kudat and elsewhere, Sabur Bonggi's economic well-being appeared headed towards a downward spiral. Because of the kinds of jobs that Sabur Bonggi tended to find themselves in when they migrated (low paying or seasonal), emigration did not revitalise the household or village economies. On the other hand, the return from the back-breaking work in mangrove harvesting was not commensurate with the hard labour investment. As well, mangroves in Sabah said to be under threat (Cabanban 2001), mangrove was banned in 2004. With this ban, Sabur reverted to growing subsistence crops more seriously. The experience of Sabur was one that has made a deep impression on Bonggi. For example, at Limbuak Darat and Batu Layar, some respondents who were actively pursuing the project of procuring title to their land were doing so from fear that if they did not have title to their lands, 'they might end up having to live in the mangroves' once development projects came in. Emigrating was an option taken up by many, and migration was equated with '*makan gaji*'. The migration patterns of Sabur Bonggi reflected the migration characteristics of the other Bonggi villages described below.

Agricultural Change and Diversification on Banggi Island

Similar to many groups in Borneo, access to land and forest are guided by an 'ethics of access' (Peluso 1996). In sum, the ethics provide for access to land to all in the 'community'. Regardless of what village Bonggi originated from, be it on Banggi or on the neighbouring Balambangan island, for as long as a person is identified as Bonggi or is married to one, s/he has access to Bonggi land. Community among Bonggi represents those who are linked by social relations (based on kinship), geography (proximity), and need (for similar findings in

6 Tausug (or Sulug as they call themselves in Sabah) assumed a culturally, and previously economically, dominant role during the ascendance of the Sulu Sultanate between the seventeenth and nineteenth centuries. Today, on Banggi, there are a few Sulug families, but among the few households, many are traders or store owners. Viewed as the one with 'religion', they own cultural capital over those who are assumed to hold on to a lesser form of folk Islam, as the Ubian, or Bajau among others. In this hierarchy, Kagayan (who are known as Jama Mapun in southern Philippines), are ranked somewhere in the middle in the Sulu system of social hierarchy. For ethnic relations of the eastern coastal zone of Borneo, see Sather (1997); for political economy of the 'Sulu Zone' see Warren (1981) and Mohd. Raduan (1995).

Ranau, Sabah, see Doolittle 2001, and in Sulawesi, see Li 1996). In the choice of land to grow rubber and fruit, in 2009, Bonggi individuals were using land that their families have earlier been allocated under the *Pertanian* scheme.

All development interventions affected Bonggi access to land, as land which they consider to be theirs through ancestral claims, being largely untitled, are officially considered to be 'stateland' and are subject to appropriation by the state for 'development' purpose. Since 1963, when Sabah became independent through merging with Malaya and Sarawak to form Malaysia, most parts of Banggi, have been visited by 'development' programmes initiated by both the government and the private sector. Prior to 1963, territorialisation, an aspect of which is exclusion, was put in place via a range of legislations involving the conversion of local control to centralised control of natural resources. Control via territorial demarcation (such as forest reserves where primacy is given to commercial timber production) via activities preferred over others (forestry over subsistence agriculture, cash crop over hill rice) is part of a whole network of legislations that exclude and prioritise rights and use (Peluso and Vandergeest 2001). In this chapter, the Sabah Land Ordinance (SLO) of 1930 is an important legislation to consider, and equally important is to note that territorialisation did not entirely succeed.

Similar to practices elsewhere such as Sarawak (Majid Cooke 2006) or Kalimantan (Eghenter 2006) legislations tend to transfer local control and management of land and forests to the state. The SLO declared all land, unless otherwise titled, as state land. For indigenous peoples, the potential for getting title to the land that they live on and, in some instances, protect, is provided for in Part 4 of the Ordinance, from Sections 64 to 86 as well as in Section 15 which lists seven different ways of claiming customary rights, and in Section 76 which covers provisions concerning communal title. However, there are two major shortcomings in the SLO to indigenous claims to land. First, fallow lands which are integral to the swidden cycle are not captured by the Ordinance – a big administrative oversight which had implications for access to land by swidden practitioners. Second, in accordance with the SLO, in applying for land under Native Title, indigenous groups are not tied down by locality; this means that an indigenous person can apply for land anywhere in Sabah, even if s/he may not originate from that locality. These shortcomings can be an opportunity for some and a problem for others. This issue is discussed later in the chapter. The main point to be raised here is that these legislations formed part of a process of centralising control over natural resources and of determining access rights to them in conformity with the state's administrative and political agendas (Peluso and Vandergeest 2001). In Sabah, as elsewhere in Malaysia, party agendas shape the form and implementation of development projects (Cramb 2007).

The Pertanian Scheme, Livelihood Diversification and Exclusion

As noted earlier, livelihood diversification is an important aspect of deagrarianisation. Deagrarianisation is supposed to be taking place when

Table 7.1 List of Government- and Private Sector-initiated Development Interventions on Banggi (1960s–2007)

Year	Type of Development Intervention	Development Agents	Area/Villages/Activities Involved/Affected	Status
Early 1960s	Agricultural and Resettlement Schemes	Department of Agriculture who implemented the *Pertanian* Scheme	Objective: Poverty alleviation. Small-holder coconut cultivation and small amount of commercial vegetables and corn. All four Bonggi settlements involved except Sabur. Resettlement of Bonggi households from dispersed hillside settlements. Infrastructure provision – schools, roads housing. Subsidies provided for the growing of coconuts.	*Pertanian* Scheme was initiated by the then ruling party United Sabah National Organisation (USNO). Bonggi were allocated land, which were untitled, 6 hectares per family at Limbuak Darat all the way to Kalangkaman and Kapitangan.
Late 1960s	*Pertanian* Scheme extended to Batu Layar		Same activities as above. Resettlement of Bonggi from nearby hills to Batu Layar.	
1974/75	Administration changed hands. *Pertanian* scheme at Batu Layar taken over by Sabah Land Development Board (SLDB), a government linked agency	SLDB	SLDB area extended further south from Limbuak at Batu Layar and inland towards the southeastern side of the island. Size of the Lok Tohog scheme —5,000 hectares. It did not include land under the Limbuak Darat *Pertanian* scheme. Logging to clear land for coconuts, experiments with oil palm. Land found to be unsuitable for oil palm.	Due to change in political party in government in the mid 1970s (from USNO to Berjaya), development of Banggi Island was not a priority and attention turned elsewhere. All land applications were frozen. Bonggi land under the *Pertanian* scheme remained untitled.

Year	Type of Development Intervention	Development Agents	Area/Villages/Activities Involved/Affected	Status
1980s	Administration changed hands. SLDB taken over by Borneo Samudra Sendirian Berhad (BSSB), a government linked agency	BSSB	No activities. Attempted oil palm. Mid 1980s land freeze broken	Funds were unavailable to support the project. Change of government – Berjaya to Parti Bersatu Sabah (PBS). Some Bonggi started applying for land title, but many did not, remembering promises of security by earlier political leaders, now deceased.
1980s	Cattle farming (ranch) introduced.	Private company	4,000 hectares required for cattle farm on so-called State land. Bonggi at Sabur had to relocate further North to the mangrove forests. Some families remaining in the Limbuak Hills relocated to Batu Layar. Extensive logging to make way for cattle ranch. Some Bonggi worked as labourers.	Labour for logging largely from outside (Sulawesi and Sarawak). Management of cattle farm was foreign. Bonggi land under the *Pertaniam* scheme remained untitled. Cattle ranch abandoned by management a few years later.
2007 to 2011	Plantation agriculture introduced.	Federal Land Consolidation and Rehabilitation Authority (FELCRA)	Rubber plantations introduced on Banggi on former SLDB Lok Tohog scheme (5,000 hectares).	Bonggi land under the *Pertanian* scheme untitled until today. In 2011 a few families obtained NativeTitles on land formerly set aside for a government agricultural station at Limbuak Darat, Some Bonggi are FELCRA participants.

Source: Interviews, Limbuk, Darat and Batu Layar (11 -13 August 2004, 1 —2 January 2008); and newspapers, Sabah Times 8 May 2006, Daily Express 9 August 2007 & 12 August 2007.

livelihoods have diversified spatially and sectorally. This section of the chapter outlines the diversification of Bonggi livelihoods under the Pertanian scheme since the late 1960s until 2007.

Table 7.1 refers to the waves of development projects that have swept across the Banggi landscape. Diversification of livelihood for all Bonggi villages under the first government-initiated project, the Pertanian scheme, was dramatic. This scheme was successful in transforming the Bonggi landscape from one of largely old growth and secondary forest (as a result of shifting cultivation) to a sea of coconuts. Bonggi were encouraged to resettle in villages along the road at Limbuak Darat and later, Batu Layar, and many complied

Coconut cultivation was successful in the four sampled villages, but that of vegetable and corn at Kapitangan and Kalangkaman was less successful being situated further north and not being accessible until the early 2000. Village vegetable production was stunted by a lack of demand and consequently, production was only confined to household use.

Diversification of livelihoods went apace at Limbuak Darat when the cattle farm was introduced and activities associated with it (such as logging) were carried out. Several Bonggi families took the opportunity to work for the cattle farm and prior to that in the logging camps, as forests were cleared to make way for the ranch.

When political parties lost interest and subsidies for the *Pertanian* scheme dried up, Bonggi persisted with their coconut production. Until recently the tenure insecurity remained unresolved . Until today, only a few Bonggi have title to the land (*Sabah Times* 8 May 2006; interviews in Limbuak Darat and Batu Layar, 11–13 August 2004). 'The issue for Bonggi today is land' (interview, former state assembly representative for Banggi Island, Kota Kinabalu, 5 December 2009). In 2011 Native Titles were issued over land occupied by descendants of participants of the *Pertanian* scheme at Limbuak Darat (personal communication, Encik Umpil November 2011).

Land development on Banggi follows practices elsewhere in Southeast Asia of viewing 'interior' people as backward, in need of 'civilising' measures, as well as appropriation of land that are categorised either as state land or unowned forests (Li 2007: 77–79). Elsewhere, it has been pointed out that much of 'state land' can be untitled customary land (Majid Cooke 2007). One point made by Li (1999, 2007) that may be useful for understanding Bonggi marginalisation is that the process must be understood in terms of the history of political, economic, and social relationships of the interior, along with modern processes of state formation (both colonial and postcolonial) and capitalist expansion.

As Table 7.1 shows, in the 1960s, the *Pertanian* scheme was backed by the support of the political party in power, the United Sabah National Organisation (USNO). When political parties in power changed, so did development agendas. The political party in power changed from USNO to Berjaya in 1976. Since Berjaya had defeated USNO, development projects initiated by the latter were put on hold. During Berjaya's time as well, all land applications on Banggi were 'frozen' except those requested by the company linked to the then chief minister, as already mentioned. It was not a cause of worry for Bonggi much because, at that time, most of them were not aware of the importance of land titles, and many did not apply for them. Their lack of interest in applying for land titles was also reinforced by assurances from someone who they regarded as their patron, Tun Mustapha, who was the first chief minister of Sabah.

The Tun said to us, just work on your land, for as long as I live this will be Bonggi land even if you do not have title. (Interviews at Batu Layar and Limbuak Darat, 7 June 2005)[7]

Later in the 1980s, when Berjaya was defeated and the Parti Bersatu Sabah (PBS) came to power, the 'freeze' on land applications was loosened and Bonggi were encouraged to apply for land titles (interviews, Kota Kinabalu, 1 May 2008). Some did apply especially for land along the road between Batu Layar and Limbuak Darat, but many did not because they were still relying on their patron's assurance that the land was theirs for as long as he was alive.

With regard to SLDB, it being a government-linked agency,[8] the 5,000 hectares assigned to it was left unattended when governments changed hands. In sum, the area set aside for SLDB management changed from being frozen during Berjaya's time to being neglected during PBS time.

When a new coalition government was formed in 1994 dominated by the United Malays National Organisation (UMNO), another 13 years elapsed before there was a resurgence of political interest on Banggi. The idea of large-scale estate-style development for growing rubber was revived through the land development agency FELCRA. However, FELCRA being a federal land development agency may not have the legal standing nor would it have the political clout to untangle long-standing issues concerning land distribution, access, and tenure, since land is a state issue in Malaysia's federal system of government. Also, in Sabah, all decisions concerning land are made by the office of the chief minister (interviews Kota Kinabalu, 19 February 2008, 1 May 2008). Nor might FELCRA have sufficient background regarding the history of development of the island to even begin to take an interest in Bonggi land entitlements. Moreover, Bonggi do not

7 Among some Bonggi who were involved with Tun Mustapha in fighting the Japanese forces during its occupation of Sabah during the Second World War, the realisation that Tun Mustapha was not alone in that struggle, made him less of a hero than the myth that surrounded him. The younger Bonggi were told by their grandparents that there were other armed forces fighting for Banggi and Sabah generally, including Australian forces (interviews in Kalangkaman, 30 December 2007). Consequently, among some younger Bonggi the image of the Tun as patron is starting to unravel.

8 An indication of the role of SLDB in the context of Sabah development can be seen in the exposed case of it being used for providing elected officials with wealth via share transfers and for supplying political parties with election funds (*New Straits Times* 17 August 2008; *Bernama*, 20 July 2007 [Online] http://www.bernama.com accessed: 2 May 2008; *New York Times* 3 May 2008). Thus, it can be concluded that SLDB would not have had the will to find alternative paths for development on Banggi if it was not pushed by political parties in power.

have any significant political representation at the local or state level, so FELCRA may not be persuaded to treat the matter as urgent or important.

In the villages, Bonggi are aware of their marginal position and is aptly captured below:

> We know the land approval process. The Land and Survey verifies the location of the land, the Land Utilisation Committee evaluates, recommends or rejects. Our problem is that we do not have anyone who pushes our interest. There is no Bonggi representative in the state assembly, nor in the offices of the state assembly. At the local representative office, they are all Ubian. We have no representation, no voice, and no area allocated to us to live. (Atin, Kalangkaman, 30–31 December 2007)[9]

When processing their applications for Native Title, Bonggi encounter many hurdles.The descendants of the original settlers at Limbuak Darat and Batu Layar have discovered that the paperwork regarding the *Pertanian* scheme was reportedly lost, which means that there are now no official government records of the scheme's existence nor for their allocated entitlements. A few families who were resettled at Limbuak Darat further found that the plots allocated to them were, at the same time, reserved for a government agricultural station.

Other applicants at Limbuak Darat found that the area that they have applied for under Native Title provisions, are at times already applied for by outsiders (individuals or companies not from Banggi). The freezing and unfreezing of land applications, noted earlier, while detrimental to the Bonggi, allowed outsiders to apply for land during the unfreezing period. A view of the land records suggest that some of the applications were made in the 1980s coinciding with the 'unfreezing' period introduced by the PBS led government. Others have been made much later subsequent to the 1980s. By the 1990s, there were some 2,000 unprocessed applications, many from outside interests and therefore, overlapping (*permohonan bertindih*).[10]

In many instances, when checking on the status of their land applications, Bonggi found that approval for titling may not be granted until the overlapping claims were sorted out by the Land and Survey Department. This means that despite their claims to indigeneity, their established customary rights to land do not have priority, so that their applications are given the same importance as those made by other indigenous applicants from outside Banggi who may not have established rights.

---

9 For example, in the 2008 elections held in early March, the Banggi seat was fought over by five contenders, none of them Bonggi (*Daily Express* 25 February 2008, pp. 1, 2). The winning candidate was an Ubian.

10 The fact that under the Sabah Land Ordinance any Sabah indigenous persons can apply for land anywhere in the state under Native Title without having to be a resident of the place, can be considered a loophole in the law.

> We were told to go see the boss at the Department, but after we met the boss, the office people told us he was the wrong one, and we should have met with the second boss…. I think they are just playing us around … we are just like cigarettes (*sigup*) that can be thrown away any time. (Tugal, Kalangkaman, 1 January 2008)

In 2007, having been unsuccessful in getting Title through normal channels, Bonggi of Limbuak Darat tried a different avenue by approaching the Malaysian Human Rights Commission (SUHAKAM). Despite the imminent threat of losing their land, the language used in their note to SUHAKAM was reconciliatory; it was a language of rights:

> We do not want to obstruct the agricultural station from coming to our village because we have received much help in our effort to grow coconuts since 1964; but we do not want our rights to live on the land we have worked on since our grandparents' time to be ignored. If this problem is not taken into consideration, we are sure that our livelihoods will be threatened. (Letter to Suhakam, Limbuak Darat, 11 February 2007 .[11]

Equally, many at Limbuak Darat and Batu Layar have found a way of resisting potential displacement. They do so by attempting to farm cooperatively on land which are officially considered to be 'state land' located outside the original Pertanian scheme and outside the FELCRA project area. Cooperative labour among relatives and kin is organised for the purpose of growing rubber and, to a smaller extent coconut and fruit trees. Inspired by ideas of self-help emanating from religious institutions, Bonggi Christians and Muslims were among the 23 families at Limbuak Darat who were working together (*gotong royong*). According to the younger members of the group, the idea was to get the production of rubber started, and ask for government assistance later. 'We do not want to be like our parents, just waiting for government assistance and being disappointed in the end' (interviews, Limbuak Darat, 20 January 2005).

> We grow rubber first, we'll worry about land titles later. If they take our land away I will ask them 'where am I from, and where are you from?' Why are you taking this land away from us? Although we don't have title [*geran*] this is our birth place, our blood [*tanah tumpah darah*]'. (Kaun, Limbuak Darat, 1 January 2008)[12]

---

11 The plan for the government agricultural station to be located at Limbuak Darat has been shelved (personal communication, Village Head (*Ketua Kampung*) Limbuak Darat, October 2011). Nevertheless, for as long as the area remains untitled it is subject to landuse change at the discrestion of the State.

12 All names are fictitious. The Church that is most active on the ground, at Limbuak Darat and Batu Layar is the Borneo Evangelical Church or locally known as the Sidang Injil Borneo (SIB). Many Bonggi, however, are Muslims, but they have carefully manoeuvered

It is clear from their action that growing rubber on 'state land' is an attempt by Bonggi to make their claims for land based on customary rights visible.

Migration Characteristics

Taking into account the interface between non-farm work and farm employment is an important aspect of the literature on agrarian transition because remittances from off-farm work have proved to be the catalyst for many households to exit agriculture and out of poverty. In many instances, it is young people in mature households who migrate who are counted upon to contribute to household incomes through remittances (Rigg 1998: 507; 2005: 187). Or conversely, the availability of young people to work in the family farm, frees up the labour of the heads of household for non-farm work (Rigg 1998: 507). If household members earn sufficiently high income, this may certainly be true. There are new houses being built in Kalangkaman and Limbuak Darat as a result of non-farm work (field notes, Kalangkaman, 31 December 2007 and Limbuak Darat, 29 November 2009).

Over the years, many informal conversations held with young as well as older people who have returned to the island indicate that earnings are not sufficiently high for remittances to be a reliable source of income for households left behind. Data based on interviews with 24 returned migrants, used in this section of the chapter captures some of the main characteristics of non-farm work engaged in by Bonggi.

The sample, admittedly small, is nevertheless useful for supplementing impressions built on observation and conversations over the last six years. Collected randomly, the sample successfully captures those who returned from a range of age cohorts.

Brief Summary of Key Findings

Table 7.2 shows that many more who returned are engaged in farming than not. All six respondents in the 51–60 age category were engaged in farming. Those not farming (nine respondents) were from the younger age groups of 21–30 and 31–40 years, although two were farming part-time and one respondent wanted to farm in the near future. Obviously, farming is considered an acceptable activity even among the young. Those not farming were variously engaged in non-farm work including working for FELCRA, for contractors engaged in building roads and housing, or in other non-farm work such as being a security guard.

religious affiliations in ways that benefit, not divide them. 'No matter what religion we belong to, we are still relatives' (interviews, Limbuak Darat, 4 June 2004).

Table 7.2 Age by Employment Activity upon Return

Age (in years)	Farming	Not Farming	Total
21–30	2	4	6
31–40	4	3	7
41–50	2	1	3
51–60	6	0	6
60+	1	1	2
TOTAL	15	9	24

Note: N = 24

Source: Fieldwork 2009

Except for four respondents, the majority had no formal education (Table 7.3). The lack of education among the Bonggi channeled them into unskilled non-farm work (Table 7.4). Although having made an exit from subsistence agriculture, most could only find work in commercial agriculture (coconut and oil palm plantation), in commercial fishing or in forestry (logging). Jobs that are not agriculture related are factory work. In the Sabah economy, factory line jobs and those in forestry and the oil palm industry are lowly paid, often taken up by migrant labour precisely because they are shunned by many locals for being low paying (Majid Cooke 2009).

Table 7.3 Education by Gender

Educational Level	Women	Men	Total
Completed high school	1	0	1
Some high school	–	2	2
Some primary	–	1	1
No schooling	3	17	20
Total	4	20	24

Note: N = 24

Source: Fieldwork 2009

Table 7.4 Destination of Migration and Nature of Work (Multiple Responses)

Destination of Migration	Nature of Work	Total
Kota Kinabalu	Day/Casual labourer in shops, digging drains	3
Kudat, rural	Coconut and oil palm plantations	10
Kudat, rural	fishing fleet (*pukat tunda*)	8
Sabah East Coast (Sandakan, Kinabatangan, Telupid, Pitas, Lahad Datu)	Timber camp, agricultural labour, builders' labourer (contract work)	7
West Coast Sabah (Sipitang)	Fishing fleet, timber camp	2
In and out and intermittent trips Sabah Westcoast (Kota Kinabalu, Kota Marudu, Kudat)	Plantation, fishing fleet and builders' labourer	6
Peninsula Malaysia: Kuala Lumpur, Penang, Trengganu	Factory and holiday resort work	3
Singapore	Builders' labourer	1
Total number of jobs held		50

Source: Fieldwork 2009

The main reason for Bonggi migration is economic, namely, in search of paid work and for extra cash for meeting basic needs of shelter, food, and education (see Table 7.5). Generally, even after long years of being away (67 per cent of respondents went away for 10 years or more) there were very little savings (Tables 7.6 and 7.7), so that many aspirations of wanting to help family or to raise household incomes were not met.

That non-farm extra local work opened up a whole new window of opportunity for new forms of employment appeared to have benefited some Bonggi, and they seized the opportunity by finding work all over Sabah, Peninsular Malaysia, and, to a smaller extent, internationally in Singapore (as evident from Table 7.4). Nevertheless, the model of non-farm work feeding into the income of farming households, sufficient to provide impetus for further diversification of household economies, does not apply in this instance because of the inability of sending sufficient funds home due to low wages earned away from home. The lack of remittances means that the interpenetration of urban and rural incomes important in the period of transition did not occur so that in many villages, the two northern ones of Kapitangan and Kalangakaman included, the livelihoods tend to remain agricultural. By extension, despite diversification of livelihoods and income structures, the agrarian transition is not taking place, at least, for the moment.

However, small gains were made in material terms. Table 7.7 shows that from non-farm work, four respondents were able to build dwellings and two were able to get married. The one respondent who had planned on using his savings to process land title application for the land on which his family had established customary rights, in the end, gave up because of the tortuous procedure and the money being too tempting when not used became depleted. So his family land remains untitled.

Table 7.5 Reasons for Migration (Multiple Responses)

Find extra cash	10
Find work	7
Other economic: School fees, help family, raise living standards	6
Follow family	4
Buy basic needs (clothes, food)	2
Other reasons: expose children to city education, to see other places	2
Total	31

Source: Fieldwork 2009

Table 7.6 Length of Time Away

Less than a year	2
1–3 years	2
4–6 years	1
7–9 years	3
10–12 years	2
13–15 years	3
16+ years	5
In and out 10 years or less	3
In and out 11 years or more	3

Note: 67 per cent of respondents have been working/living outside Banggi for 10 or more years. Range = 8 months to 42 years.

Source: Fieldwork 2009

Table 7.7 Savings (Able to save while away?) (How to use savings?)

Yes	0	
Yes, a little	6	Usage: To build a house: 4 To get married: 2 To process land title: 1
No not at all	18	
Total	24	

Source: Fieldwork 2009

Interestingly, reasons for return are only partly economic, namely, not earning enough from non-farm employment (Table 7.8). Identity and family reasons are the major drivers for return. After being away, the land beckoned to be worked on and protected, the desire of wanting to start an independent life became strong, as was the thought of enjoying family and friends again. As noted earlier, protection of the land (*jaga tanah*) is based on the fear that if untitled and apparently unimproved, the land could be regarded as unowned 'state land', and can be taken away for 'development' reasons or encroached upon by outsiders.

Table 7.8 Reasons for Return (Multiple Responses)

Salary not enough/job insecure	5
Family: Get married, to be with family	8
To look after own land	8
Identity: To return to my own place/want to start own life/not wanting to be in wage work any more	8
To rest	2
Old age	1
Total number of reasons	32

Source: Fieldwork 2009

Table 7.9 shows uncertainty about the future in Banggi. The high level of uncertainty has to do with anxiety over land. The view of the future being the same (unchanging) refers to a level of pessimism about improvement happening in their lives. A woman, Losu from Kapitangan expressed the pessimism thus:

> Same as now. I want to improve but how? The best I can do is to farm, that's all, for our own use. Not much we can sell, transport is expensive. My baby I will

> take to the clinic only when I have the money. We have land but we don't know how to process it for getting title. I have never been to school. Depends on the headman. (Losu, Kapitangan, 29 November 2009)

Table 7.9 View of Future in Banggi

Good	0
Quite good, if we try	2
Same as now	5
Uncertain	8
Harder	2
No response	7

Source: Fieldwork 2009

Especially for those in the northern part of the island, at Kapitangan where the major livelihood activity remains agriculture and the prospect of non-farm work less certain than the opportunities presented at Batu Layar and Limbuak Darat, the experience of living in town provided respondents with a basis for comparison. In this comparison, the advantages of living with family, on one's own land, of being in control of one's own time far outweigh the opportunities associated with wage work (*makan gaji*). To return was, therefore, a calculated risk.

> It's good to come back to one's own country, even if it is not that good, it is your own. In other people's country no work, no food. When the *towkay* (the employer) does not like us, we have to leave. Here when we want to work, we work. Here we have many relatives. It is easier here in the village, because there are many relatives. (Omar, Kapitangan village, 28 November 2009)

A few respondents (six), however, were planning to return to waged work in the next year or so for the same reasons that they left the first time (largely economic); but for most of those who returned, Banggi is where home is.

Conclusion

Drawing attention to the long-term changes that have taken place in the structure of rural livelihoods and incomes have generated two kinds of responses among scholars. On the one hand, there are those who regard these changes as deeply disturbing for its potential of making people already living in poverty even poorer because of clear inequalities in access to non-farm work. The deagrarianisation approach, on the other hand, views these changes as opportunities for creating flexibility, choice, and opportunity, although there is a recognition that choice made now may in the long term produce poverty (Rigg 2006: 195).

There is a shared understanding in the deagrarianisation literature though suggesting that diversification provides the rural poor with the opportunity to survive in the village, although increasingly, not on the land. This position allows for the development of an argument that the best way to alleviate poverty is to direct resources towards expanding non-farm work, not in providing services for agriculture (subsidies, financial assistance) or in land reform . To engage in the latter could be viewed as having a negative effect as it encourages the poor to remain poor.

It may be that policies should be aimed at oiling and assisting the process of transformation of farmers into non-farmers, and rural people into urbanites, rather than shoring up the livelihoods of small holders through agricultural subsidies, land reforms and piecemeal employment creation schemes. (Rigg 2006: 1950)

What this chapter has tried to do is to ask the question: what kind of policy would the deagrarianisation approach produce for people who willingly diversify and yet remain poor because the interpenetration of urban with rural incomes and resources have not occurred? The study has shown that exit conditions characterised by marginalisation have not helped Bonggi get jobs that would hasten their way out of poverty. The good job opportunities in the non-farm sector being available do not necessarily mean they are easily accessible to groups such as Bonggi. Consequently, a practical question that has to be dealt with is: How can mechanisms be built so that access to non-farm jobs are more equitable and not be structured by exit conditions that had marginalised those leaving agriculture in the first place?

If the answer lies in providing those making an exit from agriculture with the necessary skills to equip them with the mechanisms to manage their lives better in the non-farm economies, then what is the best avenue to follow – market or state-led? Some who exit from agriculture may be able to take advantage of the assistance programmes for enhancing skills or for capacity building. What may happen down the line is the creation of jobs as land holdings consolidate, and new groups of agrarian entrepreneurs and large landowners emerged. The sociological question would then be: Does power and control that is bound to accompany such entrepreneurial activity necessarily make for responsible farm management that will decrease poverty levels? This chapter has shown that elite control of resources on Banggi Island has provided low-paying non-farm jobs, but it has also created practices that exclude indigenous Bonggi from tenure rights to customary land. Since power and control in agrarian transitions are not linear, it can be regressive and accelerated as in Sabur or slowed down as in the other Bonggi villages. In Sabur, the transition has ushered in a deeper level of poverty.

References

Banggi Sub district Administration. 2004. Profile of Banggi Island. Banggi, Kudat.

Boutin, Michael E. 1994. The sociohistorical context of English borrowings in Bonggi, Paper presented at the Borneo Research Council Third Biennial International Conference Pontianak, West Kalimantan, Indonesia, July.

Bryceson, D. 2002. The scramble in Africa: Reorienting rural livelihoods. *World Development*, 30(5), 725–39.

Bulan, and Locklear, A. 2008. *Local Perspectives on Native Customary Land Rights in Sarawak*. Malaysian Human Rights Commission (*Suruhanjaya Hak Asasi Manusia* – SUHAKAM), Kuala Lumpur.

Cabanban, Annadel. 2001. *An Overview of the Sulu Sulawesi Marine Ecoregion*, Paper prepared for the workshop to formulate the Biodiversity Vision for the Sulu-Sulawesi-Marine Ecoregion, Organised by the World Wildlife Fund the Philippines, Manila, 3–5 March.

Cramb, Robert. 2007. *Land and Longhouse: Agrarian Transformation in the Uplands of Sarawak*. Copenhagen: Nordic Institute of Asian Studies Press.

Doolittle, Amity. 2001. From village land to 'native reserve': Changes in property rights in Sabah, Malaysia, 1950–1996. *Human Ecology*, 29, 69–95.

Eghenter, Cristina. 2006. Social, environmental, and legal dimensions of *adat* and its role in conservation areas – A case study from East Kalimantan, Indonesia, in *State Communities and Forests in Contemporary Borneo*, edited by Fadzilah Majid Cooke. Canberra: Australia National University

Li, Tania Murray. 1999. Introduction, in *Transforming the Indonesian Uplands*, edited by Tania Murray Li. Singapore: Institute of Southeast Asian Studies.

Li, Tania Murray. 2007. *The Will to Improve, Governmentality, Development, and the Practice of Politics*. Durham and London: Duke University Press.

Majid Cooke, Fadzilah. 2003. *Living at the Top End: Communities and Natural Resources in the Kudat Banggi Region of Northern Sabah*, WWF Project Report no. MYS 486/03, Kota Kinabalu, Sabah.

Majid Cooke, Fadzilah (ed.). 2006. Introduction: Development and conservation interventions in the beginning of 21st century Borneo, in *State Communities and Forests in Contemporary Borneo*. Canberra: Australia National University.

Majid Cooke, Fadzilah. 2007. *Demography, Local Communities and Changing Land Use in Sabah*, Interim Report of the Sabah Land Use Project, Unpublished report submitted to the World Wildlife Fund Sabah, Kota Kinabalu.

Majid Cooke, Fadzilah. 2009. In situ off-farm work in the transport industry among oil palm smallhoders in Sabah: Negotiating the borders of licit and illegal activities. *Asia Pacific Viewpoint*, 50(1), 43–57. 50th Anniversary Special Section, Fadzilah Majid Cooke (guest editor), April.

Mohd. Raduan, Mohd. Arif. 1995. *Dari Pemungutan ke Penunda Udang: Satu Kajian Mengenai Sejarah Perkembangan Perusahaan Perikanan di Borneo Utara 1950-1990* (From Collectors to Purse Seiners of Prawn: A Historical Study of the Fishing Industry in North Borneo 1950–1990). Kuala Lumpur: Universiti Malaya.

Peluso, Nancy Lee. 1996. Fruit trees and family trees in an anthropogenic forest: Ethics of access, property zones, and environmental change in Indonesia. *Society for Comparative Study of Society and History*, 38, 510–48.

Peluso, Nancy Lee and Vandergeest, Peter. 2001. Genealogies of the political forest and customary rights in Indonesia, Malaysia and Thailand. *The Journal of Asian Studies*, 60, 761–812.

Rigg, Jonathan. 1998. Rural-urban interactions, agriculture and wealth: A Southeast Asian perspective. *Progress in Human Geography*, p. 497–522.

Rigg, Jonathan. 2005. Poverty and livelihoods after full-time farming: A Southeast Asian view. *Asia Pacific Viewpoint*, 46(2), 173–84.

Rigg, Jonathan. 2006. Land, farming, livelihoods, and poverty: Rethinking the links in the rural south. *World Development*, 24(1), 180–202.

Rigg, Jonathan and Nattapoolwat, S. 2001. Embracing the global in Thailand: Activism and pragmatism in an era of deagrarianization. *World Development*, 29, 945–60.

Sather, Clifford. 1997. *The Bajau Laut: Adaptation, History and Fate in a Maritime Fishing Society of South-eastern Sabah*. Oxford: Oxford University Press.

Warren, James. 1981. *The Sulu Zone 1978–1898*. Singapore: Singapore University Press.

Chapter 8
Movement by Coercion: The Displacement of Indigenous Children in Australia

Suneeti Rekhari

Introduction

Indigenous[1] people in Australia are not only one of the most researched people in the world (Martin 2003), but have been historically defined in many different ways by varying legislations. An analysis of over 700 Australian legislations, led McCoquodale to find 67 different definitions of Indigeniety (1986: 7). These definitions were based primarily in terms of the so-called 'proportion of blood' that Aboriginal people were meant to be categorised by, and focused on the need to surveil and control (Dodson 2003). This has now changed and a working definition, used in the Commonwealth legislation, has been established, which states: 'An Aboriginal or Torres Strait Islander is a person of Aboriginal or Torres Strait Islander descent, who identifies as an Aboriginal or Torres Strait Islander and is accepted as such by the community in which he or she now lives' (Dodson 2003: 32). Despite this definition, Dodson reminds us that Indigenous peoples must continue to subvert the hegemonic representations and definitions, and continuously create meanings of the way in which 'we relate to ourselves, to each other, and to non-Indigenous peoples' (2003: 33). It, thus, becomes important to begin this chapter by stating that there continues to be a contested and complex notion of 'indigeneity' in Australia as well as internationally. The details of this contestation go beyond the purview of this particular analysis. Rather the issue is raised here at the start as a launching point for some of the other complexities experienced by colonised Indigenous populations.

One of these complexities relates to how Indigenous people in Australia have been forced to co-exist with immigrants for over 220 years. With personal experience

1 In Australia, the term Indigenous is used to refer to both Aboriginal and Torres Strait Islander peoples. So, officially, Australia has two groups of Indigenous people. However, 'Indigenous' is not universally accepted by Australia's Aboriginal and Torres Strait Islander people. Sometimes preference is given to terms that are used locally, for examples: Murry in eastern Queensland; Nunga in South Australia; Palawa in Tasmania; Yolnugu in the northern territory—north-east Arnhem land; Koori in Victoria and New South Wales or NSW (Jonas and Langton 1994). This affirms the complexity in definition and contested notions of Indigeneity referred to in the introductory paragraph.

of this co-existence for the past ten years as an Indian immigrant in Australia, and with the recent understanding of a culture more than 60,000 years old, I will outline the forcible displacement and dispossession of Aboriginal and Torres Strait Islander peoples by the very convicts and immigrants that came to their land.

Indigenous displacement in Australia was not only from the land, but also their homes and families. This chapter cannot cover all aspects of this displacement; therefore, I will concentrate on the forcible displacement of Indigenous children from their families. This chapter will outline some early documented child removal and also the experiences of children taken away under the directions of the protection and assimilation policies. The various policy assumptions, intents, and impacts of this forcible removal on Indigenous families are well-documented (Beresford and Omaji 1998; *Bringing Them Home* 1997; Briskman 2003; Haebich 2000; Haebich and Delroy 1999; Healey 2009; Read 1998, 1999). This chapter will refer to this existing scholarship and will primarily rely on the accounts tabled at the federal parliament in the May 1997 *Bringing Them Home* report.

The discussion in this chapter focuses only on information collected from these secondary sources. The use of secondary sources (as opposed to primary ethnographic data) is a carefully considered decision. First, my aim is to repudiate and not add to academic methodologies which, as Arbon (2008: 135) states, give non-Indigenous Australian researchers a 'right to speak for us [Indigenous people]'. I do not want to operate within the same discourses that have seen Indigenous people represented by outsiders. I attempt to engage with the concept of internal displacement, using Indigenous experiences of it and letting the Indigenous subject speak for itself through already documented evidence. Second, due to my previously stated cultural positioning as an immigrant outsider in Australia, I want to critique my own ontology by acknowledging that my writings on Indigenous issues and my representations of these would be 'vastly different to the representation defined, developed and refined by an Indigenous writer' (Phillips 1997 as cited in Heiss 2007: 1). Finally, my writing in this chapter is cognizant of comments by Indigenous academics such as Huggins, Morris, and Martinello (cited in Heiss 2007: 5) who believe that non-Indigenous writers can comment on Indigenous themes when based on historical evidence and colonial literature. This is precisely the data base used in this study.

Additionally, this discussion does not claim to speak for the experiences of *all* Indigenous people in the process of displacement. In fact, my very existence as an outsider negates any claims to authenticity and authority over subject matter. What I do aim to present is a critical overview of the child-removal processes and highlight the lived experiences of forced dislocation. Before beginning this overview, it is significant to note that the concept of displacement itself has long involved a large association with refugees or people affected by violence and armed conflict and civil unrest in poorer and developing nations (Denov and Campbell 2002). However, the forced displacement of Indigenous people in developed countries, such as Australia, has remained a little researched area. This chapter aims to address this lacuna and is situated in this volume, overall, as an examination of Indigenous displacement in Australia and the factors underpinning it, thereby, bringing it to the attention of an international readership. It aims to provide a tool for further analysis and questioning on the complexities involved in the 'movement by coercion' of a marginalised section of the Australian population.

Beginnings

The physical displacement of Indigenous people began with the legal fiction of *terra nullius*, which allowed the British to claim dominion over Australia in 1770, later affirmed by the arrival of the First Fleet in 1788. As a consequence, Indigenous people all over Australia were dispossessed from their land even before they set eyes on any settlers, as the doctrine covered all Australian territory irrespective of traditional and complex Indigenous connections to the land. Terra nullius predicated Australia to be a land owned by no one, a waste land, and most importantly as a land whose traditional owners were denied legal recognition of ever having occupied. This justification made the taking of land by settlers palatable. While it is true that some early settlers expressed concern over the rapidly expanding colonial empire in Australia and the methods employed to inhabit it, by the 1850s, the certainty of displacement and dispossession was apparent: '[A]s our flocks and herds and population increase, our corresponding increase of space is required, the natural owners of the soil are thrust back without treaty, bargain or apology' (Mundy 1857 as cited in Reynolds 1989: 76). Confusion and violent frontier battles often characterised these early years of colonisation and forced evictions from land. Seen in this light, the United Nations Educational, Scientific and Cultural Organisation (UNESCO) definition of displacement, '... the forced movement of people from their locality or environment and occupational activities' with its most common causes being 'armed conflict, natural disasters, famine, development and economic changes' (see www.unesco.org) can be related to the early colonial experiences of Indigenous people in Australia. Early colonial development was dependent on land, food, and water sources, which played a major contributory factor in the displacement or 'forced movement' of Indigenous people. The brutality of armed conflict and Indigenous resistance to European land grabs in the initial years of the formation of the colony played an additional role (Reynolds 2007). Thus, early land dispossession was necessarily characterised by violence and coercion.

Even in this early colonial atmosphere of land grabs, beginnings of the removal of children are visible. The *Bringing Them Home* report states that Indigenous children were being forcibly separated from their families since the very early days of European occupation (1997: 27). It is difficult to estimate the exact number of children who were displaced either at this time or later because of improper or no record-keeping; some records were lost while some were destroyed. However, some existing records reveal that in late 1788, Governor Arthur Philip captured an Aboriginal man, Arabanoo, to set-up communications between the settlers and

Aboriginal tribes in the area (Tench 1979: 140–51) around the newly established penal colony in the Sydney region of NSW. After this 'experiment' of Philip proved mildly successful, Surgeon General John White and Reverend Richard Johnson adopted two Aboriginal children, Nanbaree (a nine-year-old boy) and Boorong (a twelve-year-old girl) into their own families. The expectation was that these children could be educated, 'civilised', and taught to be bilingual to facilitate a 'more intimate and friendly intercourse with the Aboriginal community' (Hunter 1968 [1793]: 94). Nanbaree and Boorong had been orphaned in a small-pox epidemic that ravaged the Sydney Aboriginal population in 1789. They were, thus, not forcibly removed from their families. Nonetheless their relocation was not made to other members of their extended kin group or tribal group, but was carried out for and by European members of the emerging Sydney colony.

In 1814 Governor Macquarie set a historical precedence and founded the first Native Institution for Aboriginal children at Parramatta, near Sydney in NSW. It was primarily set up to take children from their families to be educated and 'civilised', with an annual feast held each year to 'reunite' parents with their children and develop better relations with Aboriginal people there (NSW Department of Aboriginal Affairs 1998: 10). Other examples begin to detail further processes of removal of children. In Western Australia, one of the first recorded instances of child removal (detailed by Haebich and Delroy 1999) took place in 1833 when Lt Governor Irwin executed the father of an eight-year-old Aboriginal boy named Billy and took the child with him to live in Fremantle. Irwin writes:

> The child has been kept in ignorance of his father's fate and it is my present intention to retain him in confinement and by kind treatment I am in hopes from his tender age, he may be [accustomed] to civilised habits, as to make it improbable he would revert to a barbarous life when grown up (Quoted in Haebich and Delroy 1999: 8).

While the removal of children from their families was not, at this time, formalised or legalised, little was done by the settlers to stop it.

Initially, the missionaries and other 'philanthropic' Europeans believed that the key to success in civilising the 'savages' was through their children (Reynolds 1989: 170). Indigenous children were targeted early on for removal from 'tribal influences'. It was at the missions that they could be trained, educated, and taught Christianity. However, their ultimate motives were to 'inculcate European values and work habits in children, who would then be employed in service to the colonial masters' (Ramsland 1986 as cited in *Bringing Them Home* 1997: 2). At missions, Indigenous people were 'somehow expected to live highly moral and upright lives' and their existence became the 'subject of gossip and analysis among government administrators and church officials' (MacFarlane and Deverall 1993: 70). Thus, Indigenous families became captives of government and church sanctioned regulations and legislations. In the middle and later decades of the twentieth century, removal to missions was especially applied to children who were considered *half-caste* or of mixed blood. This was due to the misguided belief that Aboriginal children of mixed descent were intellectually and morally superior to *full blood* Aboriginal people.

> The half-cast is intellectually above the aborigine [sic], and it is the duty of the State that they be given a chance to lead a better and purer life than their brothers. I would not hesitate for one moment to separate any half-cast from its aboriginal [sic] mother, no matter how frantic her momentary grief may be at the time. They soon forget their offspring (James Isdell 1908 as cited in Haebich and Delroy 1999: 18).

The imprint of 'blood' on these children, and the government and widespread marking of it, greatly influenced their future. Many colonists believed that Indigenous children should be trained in missions so that they could take their place in society as workers. Forcible removal also worked hand-in-hand with the negative stereotype of Aboriginal women as bad and neglectful mothers who would 'soon forget their offspring'. However, this was far from the truth as is seen by the numerous letters and correspondences of distraught Indigenous parents seeking to contact their children. The following inquiry evidence reinforces this:

> My mum had written letters to us that were never forwarded to us. Early when we were taken she used to go into the State Children's Department in Townsville with cards and things like that. They were never forwarded onto us (Confidential evidence 401, Queensland, *Bringing Them Home* 1997).

The perceived inability or failure of Indigenous parents to look after their children was often used as an excuse for the children's removal. As the evidence cited suggests, children were not told about their families and parents trying to contact them. Eventually, through legislation, the reasoning of neglect was not needed at all, even though the belief was held in the wider community. Ultimately, what began as land dispossession turned to the later cumulative effect of family and community disintegration. How this displacement was facilitated by singular government efforts through legislation is discussed next. These government measures clearly point to the coercive displacement of Indigenous children in Australia.

Coercion by Legislation

By the 1880s, residents in country towns had started demanding that the state take greater control over the Indigenous population that had been displaced by European colonisation (Read 1983: 61). The government proposed the establishment of a protectorate system so that Indigenous communities would not interfere with colonial land claims. However, by the middle of the nineteenth century it was evident that the protectorate system had failed. The government's response to this was to assign responsibility for the welfare of Indigenous people to a Chief Protector or Protection Board (*Bringing Them Home* 1997: 28). A 'Protector of Aborigines' was appointed in 1882 by the NSW government following agitation from a non-government body called the Association for the Protection of Aborigines, which wanted to ameliorate 'the present deplorable condition of the remnants of the Aboriginal tribes' (NSW Department of Aboriginal Affairs 1998: 30). In 1883, the government established a Board for the Protection of Aborigines, which was not set up under any legislation and passed many protection policies unsanctioned by law. Along with its many other policies carried out in the name of so-called 'Aboriginal protection', it practiced the removal of Aboriginal children from their families, ostensibly for the sake of the education and 'protection' of these children (Read 1981).

At this juncture, it is important to point out that a multitude of policies come under the rubric of 'protection and assimilation'. There was also no unified approach by the various state governments over the so-called 'Aboriginal problem'. Indigenous people came under the responsibility of state governments till the 1970s despite repeated calls for federal control (Haebich 2000: 161). However, there were general trends in policies and practices as outlined in the various legislations that were followed by the state governments. Some of these are highlighted ahead, especially those related to child separations.

The removal of children to 'training homes' and 'educational institutions' remained unlawful till the passing of the Aborigines Protection Act of 1909, which gave the Aborigines Protection Boards legal authority to carry out the removal of such children. With the passing of this Act, child removal and displacement became legitimate. Section 7c of the Act gave the Board the duty to 'provide for the custody, maintenance and education of the children of aborigines [sic]' (New South Wales Statutes, Act Number 25, Aborigines Protection, 1909, p. 145). Section 13 of the Act states:

> Any person who entices a child apprenticed as aforesaid to leave his lawful service, or who entices the child of any aborigine [sic], or of any person apparently having an admixture of aboriginal [sic] blood in his veins, to leave any school, home, or institution, without the consent of the board, shall be guilty of an offence against this Act (New South Wales Statutes, Act Number 25, Aborigines Protection, 1909, p. 147).

Section 16 of the Act allowed the Board to go to the children's court to seek maintenance cost of an Indigenous child aged between 5 and 16 years from near relatives. It allowed the Board to seek future maintenance or even 'past maintenance of such child, whether such child be alive or not at the time of the application' (New South Wales Statutes, Act Number 25, Aborigines Protection, 1909, p. 147). By 1911, the Northern Territory and every state except Tasmania had legislation giving the Chief Protector and the Protection Board extensive power over Indigenous people (*Bringing Them Home* 1998: 28).

Despite the extensive powers that the Board was able to exercise through the Aborigines Protection Act, they were still unhappy with it and pressed for amendments which were made in 1915. The most significant amendment that the members of the Board pushed for and were granted was the power to remove any child without parental consent or a court order. New sections that were inserted in the Act specifically allowed for this; Section 11A allowed for every child who was apprenticed, but who refused to go to the person to whom the Board had apprenticed him to, may be removed 'for the purpose of being trained, to some home or institution as the Board may arrange' (New South Wales Statutes, Aborigines Protection Amending Act, 1915, p. 122). Section 13A also allowed for the Board to 'assume full control and custody of the child of any aborigine [sic], if after due inquiry it is satisfied that such a course is in the interest of the moral or physical welfare of such child' (New South Wales Statutes, Aborigines Protection Amending Act, 1915, p. 122). This Section did allow for the parents of a child who was removed to appeal to the children's court: 'The parents of any such child so removed may appeal against any such action on the part of the Board to a Court as defined in the *Neglected Children and Juvenile Offenders Act, 1905*, in a manner to be prescribed by regulations' (New South Wales Statutes, Aborigines Protection Amending Act, 1915, p. 122). However, these courts were usually located in urban areas which were far away from where most Indigenous families lived. With little or no access to a formal education and speaking languages other than English, it would have been virtually impossible for the parents to successfully raise an appeal.

Further amendments to the Act in 1918 extended the powers of the Board and changed the definition of Aboriginality to include half-caste natives. This was in response to the perceived 'problem' of part-Indigenous people increasing in number (NSW Department of Aboriginal Affairs 1998: 47). In 1882 only 27 per cent of the known Indigenous population had been of apparent mixed descent, but by 1900 this proportion had arisen to 55 per cent (Smith 1975: 140). The Aborigines Protection Board saw its charges as posing a very real cultural as well as biological threat to the emerging nation of Australia and considered them

> … an increasing danger, because although there are only a few full-blooded Aborigines left, there are 6,000 of the mixed blood [Aboriginal people] growing up. It is a danger to us to have a people like that among us, looking upon our institutions with eyes different from our own (Scobie 1915: 1967).

The solution to this 'problem' was to blend away any signs of Aboriginality, especially from half-caste children. Children of mixed descent were looked upon as being in a state of 'racial and cultural limbo' (Haebich 1988: 48) and the government saw a need to address the growing mixed descent population. The aim of child removal was ultimately to separate 'full-bloods' from the 'half-castes', curb Indigenous reproduction (girls being especially targeted for removal), provide a cheap source of labour, and facilitate the 'Christianising' of the Indigenous population (Gardiner-Garden 1999) so that Indigenous children could be easily absorbed into the wider Australian society.

This belief on 'merging' and 'absorption' (identified in *Bringing Them Home* 1997: 29–31) strengthened, and it culminated in the Assimilation Policy being adopted in the 1937 National Conference of Commonwealth and State Aboriginal Authorities. One of the strongest proponents of assimilation and its adoption by state governments was West Australian Chief Protector A.O. Neville. Historically, he was the architect of the Child Removal Policy in Western Australia and Chief Protector from 1915 to 1940. Neville was one of the most influential figures in the Australian national history and an enthusiastic proponent of eugenicist strategy of 'breeding out the colour' of Indigenous people. His vision was one of Australian society without half castes. Mirroring contemporary views of the racial superiority of the whites, Neville believed that it was better for the half castes to be made more 'white' than 'black'. These contemporary views are evident when in 1936 the Western Australian parliament passed legislation that gave Neville the power to implement his 'breeding-out' policy. The new law made European and Aboriginal sexual relations a legal offence, punishable by imprisonment.

Neville led the 1937 Commonwealth Conference on the issue of the removal of half castes and the encouragement of marriages between half-caste women and white men. He asked:

> We have the power under the act to take any child from its mother at any stage of its life.... Are we to have a population of one million blacks in the Commonwealth or are we going to merge them into our white community and eventually forget that there were any aborigines [sic] in Australia? (AAPA 1937: 10–11).

This conference represented a personal triumph for Neville, for it adopted the Western Australian policy which stated that '... the destiny of the natives of aboriginal [sic] origin, but not of the full blood, lies in their ultimate absorption by the people of the Commonwealth, and it therefore recommends that all efforts be directed to that end' (AAPA 1937: 1). Thus, Neville's long-term plan of absorption concluded that

> ... [t]he problem of the native race, including half-castes, should be dealt with on a long-range plan. We should ask ourselves what will be the position say, 50 years hence; it is not so much the position today that has to be considered...by accepting the view that ultimately the native must be absorbed into the white population of Australia (AAPA 1937: 10).

Ultimately, it was not so much a question of Indigenous people being absorbed or 'saved' as it was a question of elimination, which Neville suggested when he wrote the following: '[E]liminate the full-blood and permit the white admixture and eventually the race will become white' (Neville 1930: 31). This was the essence of Neville's subsequent arguments that science, logic, and morality were on his side, and that there was no barrier to the eventual biological and social absorption of the 'part-Aboriginal' population into the European population of Australia – a belief that was echoed throughout Australia by various state protectors (Jacobs 1990). The consequences of such a policy were obvious as children were targeted for removal to native settlements, mission dormitories, or foster homes, where they were to be 'trained' to become productive members of the wider (white) Australian society. Only a small number, 25 per cent, of the children interviewed for the *Bringing Them Home* report spent time in a single institution. More than half the people (56 per cent) who gave evidence to the inquiry had experienced multiple placements following their removal (*Bringing Them Home Community Guide* 2007). This experience is described by an inquiry respondent as follows:

> So I went through foster homes, and I never stayed in one any longer than two months.... Then you'd be moved onto the next place and it went on and on and on. That's one of the main reasons I didn't finish primary school (Confidential evidence 316, Tasmania, *Bringing Them Home* 1997).

Coercion, Movement, and Genocide

The philosophy behind the legislated removals is considered as being inherently paternalistic, where chief protectors thought that what they were doing was right. Neville and others such as Queensland's chief protector from 1914 to 1942, J.W. Bleakley, have been described as 'kindly and well meaning' (Rowley 1972: 28). However, the reality of the situation was that children were forcibly being removed from their parents and were taken to settlements such as the Moore River Native Settlement in order to 'fit them for absorption' (AAPA 1937: 11). Neville's language at the 1937 Canberra conference, in describing the success of Western Australia's methods of child removal lacked any form of compassion or even recognition of the fact that his charges had the right to be given human status (Jacobs 1986: 21). Children were often told that their families did not want them or that they had been rejected by their parents.

> I was trying to come to grips with and believe the stories they were telling me about me being an orphan, about me having no family. In other words, telling me just get up on your own two feet, no matter what your size ... and just face the world ... and in other words you don't belong to anybody and nobody belongs to you so sink or swim. And they probably didn't believe I would swim (Confidential evidence 421, Western Australia, *Bringing Them Home* 1997).

In 1933, the government appointed a Royal Commission headed by Perth magistrate M.D. Moseley to enquire into the treatment of Indigenous people after increasing allegations of mistreatment and misconduct. In response to an attack by the Moseley Royal Commission in 1934, on the government and on Neville in particular over the removal of children, Neville argued that:

> There are scores of children in the bush camps who should be taken away from whoever is looking after them and placed in a settlement ... If we are going to fit and train such children for the future they cannot be left as they are ... (Jacobs 1990: 235).

Neville asserted the case for removing children in the belief that the children's plight was a human tragedy that concerned him more deeply than any other aspect of his work. His beliefs are summed up in his statement, 'the native must be helped in spite of himself' (Neville 1947: 80). He believed that it was necessary for children to be taken away to remove them from conditions which negated their chances in the 'outside world' and that 'the end in view will justify the means employed' (Neville 1947: 80).

Unfortunately the means that were employed were similar to those used in prisons and detention camps, and were carried out to remove all traces of cultural, racial, and social identity. A respondent to the inquiry recounts this cultural elimination:

> Y'know, I can remember we used to just talk lingo. [In the Home] they used to tell us not to talk that language, that it's devil's language. And they'd wash our mouths with soap. We sorta had to sit down with Bible language all the time. So it sorta whipped out all our language that we knew (Confidential evidence 170, South Australia, *Bringing Them Home* 1997).

Peter Read initially called the children that were taken away under these policies as 'The Lost Generations', and, later called them 'The Stolen Generations' (Read 1981). In his writings, Read (1999: 49) asserts that 'children particularly have suffered. Missionaries, teachers, government officials … thought that children's minds were like a kind of blackboard on which the European secrets could be written.' Carmel Bird goes further and states that the removal of children from their families was a 'long-term government plan to assimilate Indigenous people into the dominant white community' to bleach 'Aboriginality from Australian society. This attempt at assimilation was nothing but a policy of systematic genocide, an attempt to wipe out a race of people' (Bird 1998: 1). The accusation of genocide was also documented in the findings of the *Bringing Them Home* report (1997), which stated that the policies of forcible separation constituted genocide within the terms of the 1948 Convention on the Prevention and Punishment of the Crime of Genocide.

The legal definition of genocide in this Convention is 'forcibly transferring children of the group to another group … with intent to destroy, in whole or in part' (cited in Tatz 2001: 16), meaning intent with bad faith although not explicitly ruling out intent with good faith. However, the ultimate intent of the deeds is irrelevant as 'the crime of genocide is committed whenever the intentional destruction of a protected group takes place' (Starkman 1984: 1). Thus, previous arguments that government protectors were 'rescuing' children from hostility and neglect by their families became tenuous, as did the paternalistic intents of protectors such as Neville. The Convention proved that genocide does not require malice, but can be misguidedly committed in the interests of a protected population (Storey 1998: 227–30). Assimilation identified a 'problem' and justified removal as its 'solution', which the *Bringing Them Home* report argued constituted genocide. Forcible removal was not only intended to absorb, merge, and assimilate Aboriginal children, but as Tatz (2001) argues, showed a clear connection to genocide committed on other communities. Historians have argued that genocide occurred from the very early days of the process of absorbing Aboriginal people into the general population. For example, removal was formalised in Victoria in 1886 by the passing of the Aborigines Protection Act, which 'could be construed as an attempt at legal genocide. Certainly it was aimed at removing the Aborigines as a distinct and observable group with its own culture and way of life' (Christie 1979: 205).

Consequences of Displacement

As a result of the complexities discussed so far, and highlighted by the *Bringing Them Home* report and numerous historical evidence, Indigenous displacement in Australia cannot be seen as something produced solely by a single event. Certainly, the single event that catalysed displacement and land theft, colonisation, is clear. However, multiple series of colonising events after land usurpation have had an equal impact. All varied definitions (as elaborated in Crisp 2006; Stavropoulou 1998) with regards to the concept of displacement, including forced evictions, involuntary movements, and internal displacement, can be applied to Indigenous communities in Australia. Despite the differing terminology, these can be seen to refer to the same phenomenon (Stavropoulou 1998: 516), that of forcible internal displacement. Indigenous communities have been coerced to relocate, without any consultation or regard of their cultural, familial, social traditions, and bonds. Additionally, the forced removal of children from their families has led to disastrous consequences for Indigenous communities.

Traditional values and culture were radically impacted due to the absence of familial bonds and displacement of children from their lands, language, and heritage. The policies relating to forcible separations caused inter-generational trauma which has filtered down through the ages and affects Indigenous people even today.

> There's things in my life that I haven't dealt with and I've passed them on to my children. Gone to pieces. Anxiety attacks. I've passed this on to my kids … how do you deal with it? How do you sit down and go through all those years of abuse? Somehow I'm passing down negativity to my kids (Confidential evidence, 284, South Australia, *Bringing Them Home* 1997).

Displaced children grew up as 'institutionalised adults' which has led to associated family dysfunction and domestic violence, substance abuse, criminal offences, further institutionalisation, and imprisonment (Haebich and Delroy 1999: 53). This leads to continuing cycles of violence, abuse, and criminal activity for the most vulnerable members of a community – its children. Thus, even today, Indigenous children are experiencing the effects of colonialism and displacement, as poignantly elaborated here:

> As a child I had no mother's arms to hold me. No father to lead me into the world. Us taken-away kids only had each other. All of us damaged and too young to know what to do. We had strangers standing over us. Some were nice and did the best they could. But many were just cruel nasty types. We were flogged often. We learnt to shut up and keep our eyes to the ground, for fear of being singled out and punished. We lived in dread of being sent away again where we could be even worse off. Many of us grew up hard and tough. Others were explosive and angry. A lot grew up just struggling to cope at all. They found their peace in other institutions or alcohol. Most of us learnt how to occupy a small space and avoid anything that looked like trouble. We had few ideas about relationships. No one showed us how to be lovers or parents. How to feel safe loving someone when that risked them being taken away and leaving us alone again. Everyone and everything we loved was taken away from us kids (Alec Kruger in Kruger and Waterford 2007).

Indigenous inequality in social, economic, and political life is also evident. Looking at important social indicators such as health in Indigenous communities shows high rates of malnutrition as a result of poverty – high rates of trachoma, tuberculosis, ear and dental diseases, and diabetes. The Australian Indigenous population still has a 13–14 year shorter life expectancy than Indigenous people in Canada, and is lower than the Maori in New Zealand and the US Indigenous populations (Grbich 2004). Other social indicators, such as levels of criminal activity show that Indigenous people are more likely to be convicted of crimes, especially small offences, than non-Indigenous people (Cunneen 2001). They are also more likely to be imprisoned for longer periods. Indigenous juveniles are more likely to be arrested, referred to court and sentenced to imprisonment than non-Indigenous youth (Veld and Taylor 2005). These statistics are un-proportionately high in other areas such as suicide, deaths in custody, unemployment, alcohol, and substance abuse. Displacement has, thus, led to an avalanche-like effect on social problems and structural inequality for Indigenous Australians.

Conclusively, displacement has not only led to a loss of Indigenous cultural knowledge, but also in many cases a sense of fragmentation of identity, which is only now slowly being regained.

> The return to community for me and my family is vital, my mother was first generation removed from country for a supposed better life. As [a] senior

> woman I have a duty to re-claim what was denied. I also have an obligation to the future generation to present an alternative to mainstream [society] to ensure the survival of culture (Cummings 1993: 59).

The alienation from culture felt by generations of removed Indigenous children has been one of the biggest casualties in their process of displacement. Today, organisations such as Link-Up in NSW offer a chance for families to be reunited. This is one of the avenues by which identification of forced removal and its consequences for Indigenous people and recognition of the 'movement by coercion' can begin to occur in the wider Australian and international communities. Only then can, perhaps, the process of healing begin.

References

Aboriginal Affairs Planning Authority (AAPA). 1937. *Aboriginal Welfare, The Initial Conference of Commonwealth and State Aboriginal Authorities*, Report of Conference, Canberra, 21–23 April.

Arbon, Veronica. 2008. Knowing from where? in *History, Politics and Knowledge: Essays in Australian Indigenous Studies*, edited by Andrew Gunstone. Nth. Melbourne, Vic: Australian Scholarly Publishing, 134–46.

Beresford, Quentin and Omaji, Paul. 1998. *Our State of Mind: Racial Planning and the Stolen Generations*. Fremantle, WA: Fremantle Arts Centre Press.

Bird, Carmel. 1998. *The Stolen Generation: Their Stories*. Sydney: Random House.

Bringing Them Home: Report of the National Inquiry into the Separation of Aboriginal and Torres Strait Islander Children from Their Families. 1997. Report commissioned by Ronald Wilson, Human Rights and Equal Opportunity Commission, Sydney.

Bringing Them Home Community Guide. 2007. *Update, A Community Guide to the Findings and Recommendations of the National Inquiry into the Separation of Aboriginal and Torres Strait Islander Children From Their Families*. [Online].

Available at: http://www.humanrights.gov.au/education/bth/community_guide/index.html [accessed: 26 November 2008].

Briskman, Linda. 2003. *The Black Grapevine: Aboriginal Activism and the Stolen Generations*. Annandale, NSW: Federation Press.

Christie, M.F. 1979. *Aborigines in Colonial Victoria 1835–86*. Sydney: Sydney University Press.

Crisp, J. 2006. *Forced Displacement in Africa; Dimensions, Difficulties and Policy Directions*. Geneva: UNHCR.

Cummings, Barbara. 1993. Claiming our future. *Family Matters*, (35, August), 58–59.

Cunneen, Chris. 2001. *Conflict, Politics and Crime*. Sydney: Allen and Unwin.

Denov, M. and Campbell, K. 2002. Casualties of Aboriginal displacement in Canada: Children at risk among the Innu of Labrador. *Refuge* [Online]. Available at: http://www.articlearchives.com/international-relations/national-security/1107679-1.html [accessed: 30 November 2008].

Dodson, M. 2003. The end in the beginning: Re-(de)fining Aboriginality, in *Backlines: Contemporary Critical Writing by Indigenous Australians*, edited by M. Grossman. Carlton: Melbourne University Press, 25–42.

Gardiner-Garden, John. 1999. *From Dispossession to Reconciliation*. [Online: Social Policy Group, Parliamentary Library, Parliament of Australia].
Available at: http://www.aph.gov.au/library/pubs/rp/1998-99/99rp27.htm
[accessed: 24 November 2008].

Grbich, Carol. 2004. *Health in Australia: Sociological Concepts and Issues*. Sydney: Pearson Education Australia.

Haebich, Anna. 1988. *For Their Own Good: Aborigines and Government in the South West of Western Australia 1900–1940*. Nedlands, WA: UWA Press.

Haebich, Anna. 2000. *Broken Circles: Fragmenting Indigenous Families 1800–2000*. Fremantle, WA: Fremantle Arts Centre Press.

Haebich, Anna and Delroy, Ann. 1999. *The Stolen Generations: Separation of Aboriginal Children from Their Families*. Perth, WA: West Australian Museum.

Healey, Justin. 2009. *Stolen Generations*. Thirroul, NSW: The Spinney Press.

Heiss, Anita. 2007. *Writing about Indigenous Australia – Some Issues to Consider and Protocols to Follow*. A discussion paper, The Australian Society of Authors [Online]. Available at: http://www.asauthors.org/lib/pdf/Heiss_Writing_About_Indigenous_Australia.pdf
[accessed: 20 February 2010].

Hunter, John. 1968 [1793]. *An Historical Journal of Events at Sydney and at Sea 1787–1792*. Sydney: Angus & Robertson.

Jacobs, Pat. 1986. Science and veiled assumptions: Miscegenation in W.A. 1930–1937. *Australian Aboriginal Studies*, (2), 15–23.

Jacobs, Pat. 1990. *Mister Neville: A Biography*. WA: Freemantle Arts Centre Press.

Jonas, B. and Langton, M. 1994. *The Little Red, Yellow and Black (and Green and Blue and White Book): A Short Guide to Indigenous Australia*. Darwin: Australian Institute of Aboriginal and Torres Strait Islander Studies.

Kruger, Alec and Waterford, Gerard. 2007. Us taken-away kids, in *Alone on the Soaks – The Life and Times of Alec Kruger*. Alice Springs: IAD Press.

Martin, Karen. 2003. Ways of knowing, ways of being and ways of doing: A theoretical framework and methods for Indigenous and Indigenist re-search. *Journal of Australian Studies*, 27(76), 203–14.

McCorquodale, J.C. 1986. The legal classification of race in Australia. *Aboriginal History*, 10(1), 7–24.

MacFarlane, Ian and Deverall, Myrna. 1993. *My Heart Is Breaking: A Joint Guide to Records about Aboriginal People in the Public Record Office of Victoria and the Australian Archives*. Victorian Regional Office, Canberra: Australian Government Publishing Service.

New South Wales (NSW) Department of Aboriginal Affairs. 1998. *Securing the Truth: NSW Government Submission to the Human Rights and Equal Opportunity Commission Inquiry into the Separation of Aboriginal and Torres Strait Islander Children from their Families*. Sydney: NSW Department of Aboriginal Affairs.

Neville, A.O. 1930. *West Australian* newspaper, 18 June, State Library of West Australia.

Neville, A.O. 1947. *Australia's Coloured Minority: Its Place in the Community*. Sydney: Currawong Publishing Co.

New South Wales Statutes. 1909. *Act Number 25: Aborigines Protection*. Sydney: Government Printer.

New South Wales Statutes. 1915. *Aborigines Protection Amending Act*. Sydney: Government Printer.

Read, Peter. 1981. *The Stolen Generations*. Sydney: New South Wales Government Printer.

Read, Peter. 1983. *Link-Up*. Canberra: Link-Up (NSW) Aboriginal Corporation.

Read, Peter. 1998. *The Stolen Generations: The Removal of Aboriginal Children in New South Wales, 1883 to 1969*. Sydney: NSW Department of Aboriginal Affairs.

Read, Peter. 1999. *A Rape of the Soul So Profound: The Return of the Stolen Generations*. Sydney: Allen & Unwin.

Reynolds, Henry. 1989. *Dispossession: Black Australians and White Invaders*. Sydney: Allen & Unwin.

Reynolds, Henry. 2007. *The Other Side of the Frontier: Aboriginal Resistance to the European Invasion of Australia*. Sydney: University of New South Wales Press.

Rowley, C.D. 1972. *Outcasts in White Australia*. Middlesex, England: Penguin Books Ltd.

Scobie, Robert. 1915. *NSW Parliamentary Debates*, Vol. 57, 27 January. Sydney: Government Printer.

Smith, Len. 1975. *The Aboriginal Population of Australia*, PhD thesis, University of New South Wales, Sydney.

Starkman, Paul. 1984. Genocide and international law: Is there a cause of action? *Association of Students' International Law Society – International Law Journal*, 8(1), 1–65.

Storey, Matthew. 1998. Kruger v The Commonwealth: Does genocide require malice? *University of New South Wales Law Journal Forum – Stolen Children: From Removal to Reconciliation*, 21(1), 224–31.

Stavropoulou, M. 1998. Displacement and human rights: Reflections on UN practice, *Human Rights Quarterly*, 20(3), 515–54.

Tatz, Colin. 2001. Confronting Australian genocide, *Aboriginal History*, 25, 16–36.

Tench, Watkin. 1979. *Sydney's first four years: A narrative of the expedition to Botany Bay, and a complete account of the settlement at Port Jackson 1788–1791*. [Reprint] Sydney: Library of Australian History.

Veld, M. and Taylor, N. 2005. Statistics on juvenile detention in Australia: 1981–2004. *Technical and background paper series*, 18. Canberra: Australian Institute of Criminology.

Chapter 9
Margin, Minorities, and a Political Economy of Displacement: The Chittagong Hill Tracts of Bangladesh[1]

Bokhtiar Ahmed

What are you looking at? Ah! water?
It's not! It's tears of you and I
A sea of agony swallowed thousand lives
It's Tannyabi's melancholy
Whatever was yours, whatever was mine
Everything lay under, the water that shine. [2]

Mapping Displacement

The only passenger boat to Bilaichari, a small shanty town in the remote south of the Chittagong Hill Tracts (CHT) of Bangladesh, leaves the district town of Rangamati at dawn, six days a week. Whenever I catch that boat on my way to Mahalu[3], I usually occupy a room on rooftop of the trawler, locally converted to a double-decked passenger boat. By sunrise, the boat cruises at the centre of a large section of the Kaptai lake. The breathtaking beauty of the first shiny sunbeams on the tranquil lake water always culls above lines from Felajeya Chakma's poetry in my mind; I feel as if the morning breeze whispers its melancholy to me: 'It's not! It's not water.' Yet, my job in the region had little room for poetic indulgence. Here, I was concerned with purely empirical facts about this lake that were poetically echoed in his verses. As poetry is an inseparable part of the social facts that a community mediates through various forms of texts they produce, the agony of his verses also prevail at the heart of the reality I was trying to understand through an ethnographic empiricism. For every socio-scientific chronicler in the CHT, displacement is a much compelling phenomenon to think about. Yet, very few of

1 This chapter is a preliminary outcome of my doctoral research on identity and developmental politics in the CHT, registered at the University of Wollongong.

2 Extracts from 'Kijingot Pug' Ber-dui' by Felajeya Chakma, collected in Dewan et al. (eds), 2006. English translation mine.

3 Pseudonyms have been employed for some people and places in accordance with the ethical guidelines of this research.

their tools have the capacity of reaching many unhealable rifts forced between a human being and a native place, between the self and its true vicinity.

From the very beginning of my ethnographic engagement with the CHT, I maintained a keen eye on the diverse forms and patterns of dislocation and settlements here which initially I tried to prism through a much received scholastic notion of 'displacement'. The multi-sited mode of my ethnographic inquiry allowed me to sojourn among different identity groups in the CHT and to learn their experiences of alienation from what was meant to be 'own place' for each of them. The communities included both indigenous minorities and migrant Bengalis, whose historical mobility and settlement in the CHT turned the region into a unique crossroad of two great historical cultures, a crossroad between the peripheries of the Indian and Southeast Asian civilisations or, so to say, a distinguished margin of the colonial and sovereign state-makings in these two regions. Yet, present conditions of displacement in the CHT and their essential ruthlessness overwhelmingly shadow such historical propositions. In 2005, when I began my first research in the CHT, I was facing inescapable facts, such as the CHT then had about 90,000 of 'internally displaced' families along with another 12,000 'refugee' families repatriated from India after the accord of 1997; 38,000 settler Bengali families were also considered internally displaced (Norwegian Refugee Council 2002: 110); more than 50 per cent of the indigenous population were forced to flee massacres, arbitrary detention, torture, and extra-judicial executions (Amnesty International 2000: 12). The Rangamati Hill Tracts, one of the three administrative districts of the region, alone accounted for 35,595 indigenous and 15,516 non-indigenous families identified as 'Internally Displaced Persons' or IDPs (Karmakar and Chakma 2009:18).

The stereotypical categorisation of people as 'refugees', 'IDPs', 'migrants', 'settlers', 'tribal', 'indigenous', 'Jumma',[4] or the classificatory definitions of displacements and statistical facts associated with them, appeared quite problematic as I dug further into my fieldwork. Theoretically, I had adopted an essentially broad definition of displacement for my research that primarily seeks to understand human relationship with place and experiences of disjuncture. I tried to contextualise displacement in greater historical mobility and enclosures in the region rather than confining it within specific cases like displacement caused by the Kaptai dam. As a result, one of the earliest difficulties was to articulate these specific statistical facts in the sites of my research which, in a way, punctuated particular groups of people from rest of the population, based on some spatially and temporally defined functional categories of displacement. I struggled to employ those categories for the people I was coming across. I found the stories of dispossession by far profound, complex, and intense to fit the operational categories prevalent in general sociological discourse on displacement. Even, in my early days, I failed to extricate any particular group as 'displaced' people from

4 A collective political identity among the indigenous tribes in the CHT, particularly fostered by ethnic movements in the Bangladesh era. See, van Schendel 1992a.

the 'rest' because the latter barely existed in reality. I hardly remember any family in the CHT which had not been dislocated at some point in the near past. Such difficulties provoked an epistemological inference into those deductive notions of displacement since their classificatory definitions were of obvious importance for a comparative understanding of the nature and forces of displacement in the CHT.

Throughout its history, human population moved from one place to another, either forced by natural or manmade disasters, or on own will. Incredibly profound complexity of the phenomenon has always been a challenge for the modern social sciences that engage human mobility as an object of systematic analysis. Their effort to formulate rational scholarship on mobility resulted in a set of ambivalent analytical categories. 'Mobility' itself is an all-embracing term which includes both linear movements from one place to another and circular forms of population movement within a specified geographic expanse. For contemporary social sciences 'migration' and 'displacement' have been two rudimentary categories by which, though often interchangeably, they seek to understand and explain the processes of dislocations. 'Migration' refers to the movement of people across a specified boundary, national or international, to establish a new permanent place of residence. The United Nations (UN) has even specified a standard duration of residence to define this 'permanency' (Guinness 2002: 4). The notion evokes inquiries about the causes and consequences of migration, yet, offer little connection to the factors that 'specify boundaries' that lies at the heart of the notion itself. Migration typologies largely depend on distance, duration, and causes of mobility, but barely illuminate the multiple ways in which time, space, and people are constructed by transcending political and economic systems. Most dominant epistemology of the migration is embedded in demographic facts and figures where it is perceived in relation to population size. For such propositions, population is a deduction of the number of deaths from the number of births and then again deduced by the numbers of in-or-out migrants. Perceiving migration through quasi-mathematical demographic change makes it difficult to conceptualise and measure human mobility as socio-cultural and political transformation of people and places rather than a mere demographic event.

Displacement, the other extensive category to refer to certain types of human dislocation, is epistemologically based upon individual will to move from one place to another. The notion is increasingly substituting for migration typology like 'forced migration'. It assumes sedentary living as an ideal condition that suggests its subject populations are immobile until and unless uprooted by specific economic, political, and environmental causes. The notion signifies displacement as a state of exception rather than a normative process of social transformation. With an inherent emphasis on 'willingness' of the displaced groups, displacement is often discussed in relation to process, event, or conditions like internal or external migration, exile, diaspora, asylum, resettlement, refuge, forced displacement, disaster, development, assimilation, deterritorialization, and, nowadays more significantly, modernisation or globalisation. Such scholastic discourses on forms of dislocations also assigns individuals with displaced forms of cultural or legal identities like migrants, asylum seeker, refugee, settlers, sojourners, native, indigenous, transnational, ethnic minority, expatriates, and IDPs. Since 'willingness' of the displaced is the primary ground on which displacement is distinguished from other kinds of human mobility, the agenda of human rights also becomes an obvious embodiment of the displacement discourse, both in academic studies and political rhetoric.

Although, firmly embedded in the greater context of human mobility and enclosures as a process, displacement has become a term of specific implication, mainly in relation to the legal frameworks, mitigation efforts, and human rights advocacy. In the early 1990s, when displacement within national boundaries had become an international agenda for agencies like the UN and other human rights organisations, the notion of IDP gained prominence as a relatively new category to explain forced dislocations. The definition emerged from the need to identify the population of concern and their particular needs, information management, legal implications, and, most importantly, from formulation of policies to manage and mitigate population relocation (Mooney 2005: 10). The operational definition for internal displacement had two fundamental assumptions; first, the involuntary nature of dislocation, and second, its demarcation of national boundaries recognised by international laws. The nature of dislocation in this respect is also widely seen as a state of 'exception' or 'disorder' which attributes a temporal value to the notion itself. As a form of legal or political identity, recognitions like the IDPs assume a territorially defined national identity without considering the possibilities of coexisting defiant forms of self-esteems among people. The origin of such institutional definition of the IDPs can be well-traced in UN documents, one of which defines IDPs as '[p]ersons or groups who have been forced to flee their homes suddenly or unexpectedly in large numbers, as a result of armed conflict, internal strife, systematic violations of human rights, or natural or man-made disaster, and who are within the territory of their own country' (UNCHR 1992: 17). The entire definition epistemologically correlates a range of circumstances that appear as an exception of a normative order under which people previously could sustain their belonging to a given place. The 'home' connotes the idea of a fixed residence and, therefore, displacement is understood as eviction from that particular place of residence while inherently excluding the economic and cultural relationships a community maintains with the larger environmental settings. It insists on specific causes of displacement rather than suggesting any generic context that engenders conditions of displacement. And more problematically, it takes national territory for granted, as a homogeneous whole, and, therefore, ignores unique properties of marginal locations where the state itself maintains a differently configured role. Although the consensus on these terms to define certain forms of displacement is loosely oriented, a normative scholarship is emerging from such institutional frameworks of understanding displacement.

Patterns of displacement in the CHT, however, exhibit many arbitrary characteristics that challenge the universality of such spatio-temporal assumptions and oblige us to look at habitual aspects of displacement historically embedded in a complex political friction between the state and its geopolitical margin. First, the presumption of sedentary living is problematic in a sense that it entangles legitimated private possession of specified amount of land. While it is an effective instrument to explain disjuncture in permanent mono-grain agricultural societies politically organised around the valley states, it is not necessarily a historical condition of settlement in mountainous frontiers like the CHT. Historically, the indigenous inhabitants of this region practiced a shifting mode of food production known as 'slash and burn' or 'swiddening'. In CHT, it is commonly called 'Jum' which means both a range of cultivation, foraging, and hunting activities, and the territory where they occur. As a form of agriculture, Jum is radically different from that of river valley agriculture in terms of method and technology of cultivation, land use, and socio-political basis of production. It has historically developed a distinct relationship between land and people based on an ecologically determined material culture. Recent historical scholarships on the region suggests that this particular mode of food production is better understood as a form of livelihood that is independent of stately regulation, and was politically chosen by many groups of people who historically fled state-making projects in the valleys (Scott 2009). Such anarchic mode of food production led to an incredibly complex history of circular mobility in the CHT which determined the economic, social, and cultural organisation of Jumma communities living here. The sense of place among Jumma people is more territorial rather than the idea of an enclosed neighbourhood with legitimated private possession of specific amounts of household land. The mountainous forests are indispensible part of their sense of home where they dwell considerable part of the year in shifting farming camps. If they are alienated from the forest they exploit to sustain their own livelihood, a condition that they are experiencing in large scale at present, a Jumma village may have the absolute experience of being displaced while its people still live in the same domicile. Their settlements cannot be accounted for a permanent house since cultivation practices generate a natural context of both circular mobility and a territorial sedenterism. In this respect, the presumption of permanent residency to understand their displacement becomes too simplistic and eventually foreshadows a great extent of displacement they encounter from within.

Furthermore, a wide range of displacement in the CHT does not necessarily ensue 'suddenly or unexpectedly' or 'in large numbers' at a time. Some cases may appear as an unforeseen disaster for the marginally ethnic communities here, like the Kaptai dam, political violence, or natural calamities. Yet, 'suddenness' as a distinguishing feature of displacement slips over many forms of systematic displacement that the territorial enclosure by the Bangladeshi state engenders. The Jumma communities, and a considerable part of Bengalis relocated here, live under a constant threat of displacement; many of them already with several experiences of being displaced in their lifetime. Displacement has been a fairly common part of their marginal existence for decades. In this regard, 'unexpected' or 'sudden', as defining principle of displacement, appears problematic in understanding its ingrained nature in the CHT as apparently 'sudden' causes of displacement have been so recurrent that they form a constant basis for alienation of the minority groups from their traditional land. Similarly, a greater number of displacement occurs in ingrained and diffused forms which the idea of 'large numbers' excludes from its propagation. The issue of willingness also requires an analytical reconsideration at least in two respects. First, when peoples' economic and cultural relationships are disrupted over a long period, dislocations might not reflect their actual will, which is at times completely undermined by political hegemony. Second, applying causal categories like 'push and pull factors', a typology of willingness prevalent in migration studies, trivialise the fact that both type of factors are simultaneously ensued and managed by different political agencies of the state. Willingness is a delicate notion to measure through rationalisation, particularly when it is both constructed and defined by powerful consensus that the state endorses.

Last but not least, displacement as a standardised notion presumes cartographically defined national boundaries and, thus, comply relevant legal frameworks. The national borders that today divide CHT from the adjacent two northeast Indian states and a Burmese state, was arbitrarily enforced upon its people after the decolonialisation of British India. Given the fact that boundaries were drawn with a sheer indifference to many of its peoples' historical territoriality, in the CHT, they first enclaved and then marginalised 11 distinct identity groups[5] with diasporic linkages to what had become 'foreign countries' almost overnight. While our common understanding of displacement conforms to the legitimacy of such borders, for many identity groups in the CHT, they are arbitrarily imposed and in many ways, undone by them. For the Pangkhuas, whose experience of displacement I shall mostly refer to in this chapter, the most general typology of migration or displacement, 'internal' and 'international', is irrelevant since their diasporic identity of 'Zo' people confront the borders as disjuncture rather than territorial consolidation. This, along with further dilemmas of identity politics, determines the nature of the state's presence and policies in this region which use displacement as an integral part of its enclosure, along with other coercive border-making practices. The scenario is at odds with any categorical proposition like 'internal' or 'international' since their traditional land itself had been divided and redefined by political events with no realistic connection to the way they perceived their habitat for centuries.

So, 'displacement' as a putative category may well slip over peoples' actual endurance towards uprootment due to such coherent spatio-temporality of its facts and figures. While we intend to understand it on the grounds of intangible realities like 'willingness' and 'sense of a place' or moralising discourse like 'human rights', it is often proven to be inadequate to reach many essential realms of those very human experiences. Therefore, in this chapter, I explore 'displacement' as an essential constituent of the dominant mode of political and economic relations in the CHT. While acknowledging the analytical and pragmatic implications of

5 The identity groups are Bawm, Chak, Chakma, Khio, Khumi, Lushai, Marma, Mro, Pangkhua, Tonchoingya, and Tripura.

measuring the 'facts and figures' of displacement in a given time and space, I argue for a better accomplishment of historically informed social, political, and economic relations that persistently reproduce the contexts and political necessities of displacement. In the remaining sections of this chapter, I first recount how displacement in the CHT is a historical legacy of the colonial and postcolonial state formation that enclaved distinct minority cultures in this territorial margin of the Bangladeshi state. Then, I outline how I differ from the implicit assumption that 'displacement' is essentially the temporal consequence of particular political conflict or development project and, therefore, can be effectively mitigated by moral, ethical, or legal responses to particular cases. What I offer as an alternative approach, is a generic political economy of displacement that suggests displacement is an essential constituent of the political order the state enforces on its margins to sustain specific political and economic relations and, therefore, any solution to it requires greater political consensus to become sustainable. My aim is also to problematise the extraordinary articulation of regulatory and disciplinary practices by the state/s in this politically punctuated territory, and to provide a subsequent analysis of how displacement is a necessary entailment of these practices.

State-making in the CHT: From Homeland to Dreamland

As an ethnographic site, the CHT has very few parallels when we consider its geopolitical location, linguistic and cultural diversity, and a distinguishing history of political enclosure by state powers. This small hilly terrain occupies 13,184 square kilometres of south-eastern Bangladesh, 10 per cent of its national territory which is administratively divided into three frontier districts of Bandarban, Rangamati, and Khagrachari. Historically, 11 distinct minority cultures inhabited the region along with a more recent history of dominant Bengali settlement which has been literally patronised by the state and nationalist politics for decades. Besides such political reproduction of its demography, ethnic insurgency and counter insurgency, development projects and industrialisation, massive exploitation of land and forest resources, coercive state control over settlement and mobility patterns have been essential parts of state-making in this region. Since late 1970s, the state's violent domination and territorial enclosure was confronted by indigenous resistance both in non-violent and militant forms. Although a peace accord was signed between the indigenous resistance and the government in 1997[6], CHT still remains an exclusive political territory, literally confined by security checkpoints. As a rationalised administrative form, the state's historical presence at this margin has an exceptional articulation that offers a unique context to rethink displacement as a substantial mode of political order.

---

6 For a full text of the accord, see The CHT Commission 2008, http://www.chtcommission.org/information-about-the-chittagong-hill-tracts/the-cht-peace-accord/ (Accessed 24 July 2009)

Until 1900, the CHT had a prolonged isolation from the modernisation of society and state in British India. Before the 1860s, the hill tribes had a regional autonomy in this ethnically defined territory which survived the medieval Mughal rule and the first half of British colonial rule[7]. The absence of any centralised governance, localised forms of political and social organisation, and simple methods of economic production allowed the local people to sustain basically self-governing small political entities without highly formalised political system (Mey 1996 [1980]: 18). In 1860, the British government that succeeded the East India Company after the 'Great Revolt' of 1857, enforced federal rules here by declaring it an administrative district. By the Act XXII of 1860, the former 'Carpus Mohol' became the Chittagong Hill Tracts (Kabir 1998: 13). Two reasons may be at work behind this administrative reform. First, to impose effective military control over the region since this belt including the Himalayan basin was sort of shadow colonies for the British which they needed to have full control after the 'Great Revolt'. Second, law and order was needed to be enforced to secure the newly introduced plantations like tea, cotton, and rubber including the exploitation of its abundant forest resources (Ahmed 1991: 68). In fact, like most other peripheral societies, the CHT also experienced the very first presence of the modern state in its most coercive form, driven by certain economic dreams and a political paranoia. The Act of 1860 is considered to be the earliest legislative document proclaiming political enclosure of the CHT by a modern state power.

This intervention was followed by further economic, civic, and administrative reforms. But unlike the successive nationalist governments, the British were never that enthusiastic to merge indigenous communities with its subject population in rest of India who already mounted a momentous nationalist uprise against Britain's colonial rule. The CHT Regulation Act of 1900, commonly known as the CHT Manual, somewhat restricted internal migration by setting the district office as the regulating authority of settlement and, thus, proclaimed to be a administrative manual that consolidate and localise each tribe round its chief, which suggests a policy to safeguard indigenous political formations. Still, the Act entrenched intrinsic colonial policies including: *(a)* strengthening the common policy of 'divide and rule'; *(b)* to hold stable political order and legislation which was necessary to increase revenue from resource extraction and plantation; and *(c)* to minimise the possibilities of any greater Indian nationalist uprise (Ahmed 1991: 59). The 1900 CHT Act had also divided the territory into three administrative units under hereditary chiefs: the Chakma, Bomang and Mong Circles. The Act partially recognised the customary authority of the chiefs from majority Chakma and Marma communities, but denied their claim for 'semi-independent local state'. However, throughout the colonial era, the state's vision of the territory was more economic than governmental. There are several instances in the Act of 1900 that clearly demonstrates the colonial state's perception of the territory as an economic dreamland and how its political enclosure reproduced many of its people as 'criminal

7 See Ahmed 1996; Barua 2001; Kabir 1998; Mohsin and Ahmed 1996.

tribes'. It was this Act that formally turned this region, what was the homeland of the Jumma people for centuries, into an economic dreamland for the state.

The Indian Act of 1935 declared the CHT as 'totally excluded area' which restricts political interaction of any sort with the rest of India. This Act also brought some dimensional changes in the local power structure. The CHT was now directly ruled by the federal government in Delhi instead of state government in Kolkata and the circle chiefs achieved some advisory power in policymaking at the cost of reduced local administrative power. In 1947, when the new state boundaries were drawn on the map of British India, the CHT was included in India, and Firozpur of Punjab was in Pakistan. But to prevent possible Sikh insurgency, Firozpur was brought back to India, and Pakistan was compensated with the CHT.[8] Soon after Partition, Pakistan, the state for Indian Muslims, initially implied the same perspective of its preceding colonial government. In 1955, CHT was entitled a 'special status' which brought it under direct federal rule from Karachi. It was opened up for outsiders after the military regime came into power in 1958. Despite some weak legislative restrictions on settlement, increasing influx of the settlers, mostly the poorer segment of dominant Bengali majority of East Pakistan and West Pakistani industrial investments, began to undermine the lives of indigenous people (Ahmed 1991: 67). The construction of Kaptai dam between 1959 and 1963, funded by the USAID, submerged 40 per cent of total cultivable plains in the region. About 100,000 people, mostly Chakmas, lost everything under water and out of them 60,000 were reported to be deprived of minimum compensations. Around 10,000 fled to India as refugees.

The Bangladesh constitution of 1972 was the first legislative instrument that nullified the special administrative status of the CHT. This, along with the devastation caused by Kaptai dam and other stately repressions had given rise to an armed ethnic resistance led by the Parbatya Chattagram Jana Samhati Samiti (PCJSS)[9] and its military wing 'Shanti Bahini' (quite literally, the 'Peace Army'). The Bangladesh state's primary response to what it typically calls 'insurgency' was sheer militarisation of the region. Every act of guerrilla warfare conducted by the Shanti Bahini was retaliated through massacres of adjacent villages, executions, summary justice, extermination, and genocides. Until 1997, the CHT observed routine bloodsheds which forced more than a 100,000 indigenous people to flee on the other side of the state borders. In this political climate of 'insurgency' and 'counter-insurgency' since the mid-1970s, successive governments implemented a planned resettlement of poor Bengali populations from other districts to the CHT.

8 The partition of India in 1947, in fact, decided the fate of millions of people by drawing boundaries almost on a frantic weekend what Willem van Schendel called 'Radcliffe's Fateful Line'. Sir Cyril Radcliffe, chairman of the Boundary Commission admits the fact as 'I was so rushed that I had not time to go into details' (van Schendel 2005: 39). The CHT was one of such negligible details that the formation of two big nations could hardly bother.

9 PCJSS or Peoples' Solidarity Union of the Chittagong Hill Tracts (hereafter JSS).

This resettlement programme remained the primary 'non-military' activity by the state to establish its nationalistic claim on the region. Bengalis, who amounted only 9 per cent of the total population of the CHT in 1959, shot up to 49 per cent according to the 1991 census. Most of the settlers were poor landless peasants who volunteered in the settlement programme. But now a few among them control most trades and politics, whereas majority of them live in poor conditions despite comparative advantages from state agencies.

However, while the entire scenario was a typical case of state-making in a previously less or ungoverned territory, displacement was mostly regarded and reported in numbers of victims of isolated incidents. The emphasis on the specific case and causes of displacement largely shadow the fact that the conditions of displacement originate from the legal enclosure of the territory, primarily by the Act of 1900 which altered the very nature of land possession in the CHT. The Act brought the entire territory under state ownership by designating them as reserve forest or unclassed state forest, which in effect dispelled the hill peoples' traditional right over their historical pasture. While this served as the principal means to alienate hill people from forest lands both by the colonial and postcolonial states, it was barely taken into account to understand the nature of displacement here. Since the customary land rights of the hill people were maintained through informal social consensus rather than written legal documents, such legal enclosure of their land had changed significant parts of their livelihood, settlement, and mobility into criminal acts and thus legitimised various forms of displacement in the CHT. Although the regulation marginally protected their access to those forest lands, at the same time, it created scopes to vandalise the ecological setting of the territory through industrial and commercial extraction of its natural resources. One might easily call this a distinct case of environmental colonialism, the earliest stage of stately enclosure of hitherto ungoverned territory. Over the decades, the indigenous people were denied access to increasing amounts of land. The natural vegetation was replaced by commercially appropriable timber woods and, therefore, indigenous people were barred from cultivating those lands. It also affected their traditional means of extracting forest produce including foraging and hunting practices. The environmental colonialism was followed by development interventions, militarisation, population transplantation, and, most notoriously, a resettlement programme of strategic hamletting. Undeniably, displacement was an obvious entailment of every means employed.

A Journey by Boat to Mahalu

It is not difficult for travellers with anthropological intentions to notice how localities in the CHT have been politically reproduced since the evidences are plenty and easily identifiable. As mentioned earlier, the CHT is a mountainous, rugged terrain that historically posed an administrative challenge to state powers which could barely penetrate in this region until the second half of the twentieth century.

Francis Buchanan's 1798 (van Schendel 1992b) account is a brilliant description that shows how little control colonial administration had in this region. Here, the transportation was literally confined within the course of the Karnafuli river and its four major tributaries in northern CHT, and along the course of the Sangu river in the south. These river routes were also highly subjected to seasonality as most of the tributary channels were navigable only for six months a year. As a result, the state had no effective control over most of its parts. For central and northern CHT, the Kaptai dam was the biggest leap in the history of state-making in the CHT as it not only generated 100 megawatts of electricity, but also provided with round-the-year navigation to a vast region that was previously inaccessible. It laid the necessary context for systematic displacements that were lately ushered by population transplantation and militarisation. In fact, the lake helped to reconfigure the settlement and mobility patterns in the region from a security point of view while newly emerged navigation routes served as strategic pathways of enforcing state control. Insurgency since the 1970s was a perfect excuse to first militarise the region and then coercively manoeuvre a certain geopolitical transformation that can reinforce the state's presence in the territory. Thus, the construction of the Kaptai dam not merely displaced an identified group of population; it was the bulwark to transform both settlement and mobility patterns in the CHT which systematically displaced a much greater number of populations.

The boat journey to Mahalu is a journey through this history of displacement. One can also catch a boat for Mahalu from Kaptai, a northern shire of Rangamati district where Karnafuli has been dammed. The jetty is juxtaposed with two crucial establishments, the Kaptai dam and its power plant on the right, and a military garrison on the left. It can be said without much controversy that this is where the history of modern CHT had begun. Majority of the Bengali civilian population in this small town resettled here in connection with the dam; two large paper mills were established here in the 1960s. The army garrison is located on a junction of several hills which looks like a small island in the verge of a huge water body. Mahalu is at least a half-a-day's boat journey from here, a physical distance of less than 40 kilometres. The sentry post of the garrison that stands in the backdrop of a giant wall painting illustrating a band of Bangladeshi soldiers pointing their muzzles to an identical Jumma village, serves as the first of the six security checkpoints where every boat has to stop for security screening. The route is generally known as Reingkhiong Line to indicate the flow of the Reingkhiong river which joins Karnafuli from the east at an hour's distance from Kaptai. The route is policed by military, border guards, and police checkpoints at strategic bends and routine patrol teams on 'speedboats'. Local country boats with slow one-stroke engines have to register with both military and local government and can only operate in specified routes. Any deviance from that has to be informed in the army camps beforehand.

Mir Hashem is proudly the oldest boat operator on the Kaptai to Farua Bazar route which is the remotest destination for regular passenger boats through Reingkhiong Line. His large engine boat can operate only up to Bilaichari in the lean water season from where he transfers his passengers into a small log boat for the rest of the journey. Every time I arrive to catch his 2 o'clock down-trip to Farua, he greets me with an amiable smile spreading on his sunburnt face. 'I am going to be late for home tonight,' says the Bengali man in his 40s, suggesting that he has to stop longer on security posts as quite often, being an unfamiliar passenger, I had to appear before long questioning sessions while he waited with the rest of his regular local passengers. My knowledge on the neighbourhoods we pass by in this journey mostly comes from his casual chats while steering the hull in his rear cockpit. Mir has been running his boat since he came here with his father from Noakhali, a coastal district of southern Bangladesh, after the terrible cyclone of 1970 that washed away his ancestral village. Since then, he has witnessed the making of the numerous Bengali and Jumma neighbourhoods that today stand on the banks of Reingkhiong. 'There was nothing,' he insists each time I asked him about the old days. He informs me:

> This military camp and the adjacent Bengali *para* [neighbourhood] was built after the Shanti Bahini attack on Bilaichari armed police camp; this village were set by the military for the hillmen who they brought from somewhere in the Borkol line; look at this Para, they were brought here from the Gilamoon hills when the Shanti Bahini set up a hide out there; this Bengali Para was set by the forest department's contractor….

Mir seems to be a cruising chronicler of the settlement history of Reingkhiong embankments besides being a key person in supporting their mobility towards the outside world. On those long biding hours on board, I also met people from different localities amongst whom many became familiar faces. The floating stories on 'Hashem Boat', as his business is locally known, clearly vindicate a sheer political significance of Kaptai lake. Historically, as evident from the earlier travel accounts, such as those of Buchanan, the indigenous people in the region, particularly the Zo tribes, lived well away from the seasonal navigation routes; in small villages surrounded by abundant mountainous forestland part of which they cultivated in shifting turns of four or five years. Although land was legally appropriated by the Act of 1900, the state could not establish de facto control over a vast region that lay far beyond navigation routes. Damming the Karnafuly river facilitated an extended navigability along its four major tributaries including the Reingkhiong. This primarily enabled the state to establish control over forest resources and natural vegetation of the forests was immediately replaced with commercially valuable timber woods. This displaced several riverside Jumma settlements by forcing them to move to further remote locations where the forest department was yet to penetrate. In the late 1970s, when the indigenous resistance against state hegemony turned counter hegemonic by militarising itself, the primary response from the state was a strategic hamletting of the indigenous localities. Most of the Jumma villages that one pass by in Reingkhiong were established by Bangladesh military, whose inhabitants were forced to move here from frontier mountains where they had previously sought refuge from increasing state expansions. Like military camps and their grid of checkpoints, the villages were also set in strategic locations to bring them under constant surveillance and policing. The official documents in many instances mention this mechanism in maxims like 'tribal development'.

Extended mobility and reinforced controls on it also facilitated a compatible resettlement of Bengali population in protected villages adjacent to military facilities. The military was also accused of using these poor landless peasants from plain land Bangladesh as human shields against the guerrilla attacks by the Shanti Bahini. The 1980s observed the biggest flow of transplanted Bengalis in the region who were offered government incentives for settling in the CHT. While travelling along Reingkhiong, it is not very difficult for someone to distinguish the Bengali settlements from the Jumma villages. The settlement patterns are evidently determined by the type of economic and political relationships each community developed accordingly with the ecology and political authorities. The Bengali houses are usually built on flat terrains close to the military camps and transportation. Most of them have clay walls with tin roofs standing on clay pedestals which are highly vulnerable to flash floods during the monsoon. Bengalis primarily engage in commercial fishing and forest resource extraction like timber or bamboo businesses. Some sources include the Bengali 'settlers' among the IDPs which has certain political implications, but little is known about the displacement they suffered in somewhere in the 'plain land' Bangladesh before they were relocated here. Like these communities, there are many others who live a displaced and dispossessed life outside the known definitions and figures.

However, the housing pattern of the Jummas on the other hand, is impressively compliant to the geological and environmental setting. Most of the hillmen, regardless of their specific ethnic identity, build elevated houses on hill steeps. The main structure of the house stands on strong wood columns that support the floor which in one end affixes the front yard or walkways in the hill steep with an open-ended triangular hollow underneath. Both the floor and walls are made of closely knitted bamboo canes. The hollow under the floors allows rainwater to flow through without destabilising its foundation. The space is also used to store fuelwood and raising livestock. The roofs are traditionally made of thatch. The average Jumma household consists of a small terrace that serves as the foyer of a large living room surrounded by one or two additional rooms including the kitchen and its open back terrace. The Jumma villages on the banks of the Reingkhiong, however, are fairly moderate in comparison to such ideal architecture of the Jumma houses. Majority of them has a much larger population than that of traditional Jumma villages in other parts of CHT do. Most of their roofs are made of tin rather than being thatched. Every village, almost without an exception, has a long concrete staircase which is the most visible modern structure; some of them do not have any other concrete structure except this.

The accounts I gathered on my routine journey along this route clearly suggest that most of these villages were set up by the military roughly around the 1980s.

Before that, inhabitants of these villages lived on remote hill tops where they had much easier access to Jum lands and, perhaps, a necessary distance from the stately persuasion. After the Shanti Bahini attacks on armed forces had become frequent, many of the bordering villages in less penetrable locations were clustered on the banks of Reingkhiong. The clustering also involved a political scrutiny of people and those suspected as insurgents had to flee or face persecution. The neighbourhoods emerged as part of a coercive political project that the state needed to implement to sustain its economic and political goals in the CHT. On my numerous journeys, or sitting on the terraces of the village huts, I often witnessed military speedboats patrolling in the river. Columns of armed soldiers sit on each side of the boats, pointing their muzzles to the neighbourhoods. Several times I also have confronted them on my long walks to remote villages or farming camps, or in the Jumma villages I lived; long file of heavily armed soldiers walking like an idly crawling snake through the neighbourhoods, looking at a marooned people whose language, culture, or livelihoods are barely intelligible to them.

The Pangkhua village of Mahalu is an ideal case of strategic hamletting. The village was establised by Bangladesh military in 1979. The official documents still call it the 'Pangkhua Rehabilitation Centre' though it does not offer any idea about what and where they have been rehabilitated from. It consists of several hills of an average height of 300 feet, half circled by a sharp bend of the Reingkhiong river. Since the mid-1970s, the area had been one of the earliest places where armed indigenous resistance groups began fighting the national armed forces. Surrounding areas were also known to have hideouts of both the Shanti Bahini and some cross-border guerrilla groups fighting the Indian or Burmese military on the other side of the border where both the Indian and Burmese states were executing a similar kind of enclosure as the Bangladeshi state in the CHT, and, were notably resisted by the indigenous inhabitants of the Indian state of Mizoram and the Chin state of Burma. The Reingkhiong river, though hazardously dry or notoriously turbulent depending upon seasons, is a crucial route both for military and insurgents. The establishment of the village, which officially claims to be the act of benevolence such as rehabilitation, in fact, has resulted from strategic military decisions.

Mahalu today is home to around a hundred Pangkhua families relocated from different parts of the CHT since the early 1980s. Like most other clustered settlements along the Reingkhiong, the village also has a long concrete staircase from the wharf which leads to the main trail on the hill tops. The residents are divided into three 'blocks' under three different Karbaries.[10] The village has a primary school on the left of the wharf and a two-storied building on the right that serves both as guest house and as a community hall. There are few other modern structures within the village, like two abandoned school buildings, a church, and several houses of the village's elite. The blocks are numbered according to their distance from the wharf, which is the gateway of the village and, ironically, also

10 Traditional village chief, restored by the 1900 CHT Act.

according to the economic and social status of their occupants. Block One consists of the most prosperous families who barely depend on Jum cultivation and have availed external sources of earning like paid jobs in private and public sectors. Many of them also run their own businesses and hold political influences of varied sources and descriptions. Pangkhuas at Block Two comprise of relatively poorer families who partly depend on Jum cultivation besides small trading or low-paid temporary jobs. The poorest and most dispossessed families reside in Block Three. Majority of them struggle to survive while being solely dependent upon Jum. A few of them take up casual opportunities to work as labourers for the forest department, in road construction or sometimes for the military.

The scenario was quite curious given the experiences I had in remote indigenous villages in other parts of the CHT. I have not come across any village that has so clearly defined socio-economic composition. According to population size and socio-political organization, Mahalu is in fact a combination of three different standard Jumma villages. The people at the first block are the oldest inhabitants in the village whereas those at the other two came later and are vulnerable to further displacement. The village is surrounded by Reingkhiong and Alikhiong forest reserves which inevitably make their cultivation a criminal act. Every year they have to negotiate with the forest department and, of course, the army camps, to grow food in the hills. Many families cultivate remote lands in more than a day's walk across the mountains where they have to live for the entire agricultural season. Such disruption in normative rights to land and territory forms a constant basis for systematic dislocations. Perhaps, dissociation from Jum cultivation is the most crucial and least attended forces of displacement in the CHT. Jum cultivation is a traditionally organised and highly collective process as I also have witnessed in several other villages of CHT. In an ideal condition, each year the village leaders decide which part of the surrounding forest will be cultivated and allocate each family a certain part of it. The families perform appropriate rituals on that land before they slash and burn the bush. Traditionally, they do not sell labour within their community. So, families work for each other in certain stages of cultivation like during plantation and harvest, which involve a traditional labour organisation based on group relations and gift economy. The entire process is now increasingly disrupted by external controls over land, territory, and settlement. The relocated Pangkhuas at Mahalu have lost their lineage-based labour organisation which has eventually infused paid works within the community. Such systematic alienations from traditional subsistence economy is a slow but steady process that does not result into massive displacement in a given moment but still accounts for a greater number of it.

Almost every adult at Mahalu has a memory of being displaced as the village itself was 'set' in the early 1980s. Pangkhua families from different locations were relocated here over a decade among whom many had to leave again under certain constrains. Many of them have a part of the family living in different places of the Indian state of Mizoram, and in fewer cases, in the Chin state of Burma. There is cross-border mobility, partly due to social linkages and partly for small-scale trading. The entire border in the CHT does not have any immigration posts and, therefore, such mobilities are inevitably 'illegal' as they occur without any formal documentation. Like their Bangladeshi counterpart, the Indian and Burmese authorities also highly restrict foreigners in the states of Mizoram and Chin. As a result, social mobilities among the Pangkhuas were also disrupted by the emergence of this new border in 1947 as much as their historical territoriality and livelihood. Many families at Mahalu were first displaced by the Indian security forces in Mizoram as suspected 'insurgents' during the years of Mizo resistance led by the Mizo National Front (MNF) that fought the Indian state until they won a considerable degree of autonomy in 1984. In those years, a large number of Pangkhua families had fled Mizoram across the fairly recent state border which itself was younger than many of them. In the 1980s, they were again displaced within or from the CHT under village 're-clustering'. Lom Lian Pangkhua, another good friend of mine, lived in Block Two with his family of seven members. A tenderly old man in his 70s, he shared his own account with me, of the long way his family ended up settling in this village during an informal discussion in 2005:

> [A]long with roughly 40 other families, we fled our ancestral village in Mizoram (presumably around mid-1970s) after few youth of the village were pulled off and shot by the Indian army after a battle between the army and Mizo insurgents in down town Aijawl.[11] We had no idea where to go as we Pangkhuas are hill dwellers and do not know any means other than the Jum to survive. For survival we need hills, streams, and forests. We were fleeing as far as possible from Aijawl. Some of us had relatives in different parts of the Kaptai, Sazek, Borkol, and Naniarchor.… I did not realise exactly when we crossed border. My family, including nine others, settled in a small Pangkhua village near Sazek. But the problem was scarcity of land. Many of us returned to Mizoram as the village did not have enough accessible Jum land to support all. Rest of us could manage to find small places in the hills to grow food. Then the army camps were set as soon as the Chakmas had started fighting the army in different places. One evening a young officer arrived in our village with his troops; they were very tired and angry after a day long walk across the hills. He enlisted each family and then ordered everyone to leave the village before next sunset. We begged him that we were not Chakmas and we had been peaceful. He asked about our community and finally could figure out that we belonged to a peaceful group without any association with the Chakmas. Then he ordered us to move here. For some time, we had been hearing that a Pangkhua leader was setting a village here with the help of the Bangladesh military. So, few of us went to see him. He allotted us this place in 'Block Two'… gradually we could manage the army and forest department to allow us to grow crops in the forest … but land is getting very scarce these days as we do not cultivate the same land each year; after every harvest we leave it to rest for a few years. But now many of us buy white

11 The capital city of the Indian state of Mizoram.

> fertiliser from the Bilaichori Bazar to recultivate the same land; sometimes it works and sometimes it does not and not all of us can afford it. This year we were supposed to cultivate the 'north', but the foresters say we cannot do that part now, so we are in problem.

Lom Lian's story largely resembles the settlement history of almost every Pangkhua family in Blocks Two and Three. Despite a sheer scarcity of cultivable land and forest pasture, Mahalu has some relative advantages than average Jumma villages. Their conforming role in political conflicts and living in an accessible location brought them few external economic opportunities. Many poorer families worked for road construction, first under Food for Work programme of the Bangladesh government and, lately, under a food security programme supported by the World Food Program. It is one of the fortunate villages that had, though interrupted, a primary school throughout its existence. It is also easier to send children to high school or colleges at Rangamati from here. People can also avail temporary employment in various private and public interventions. Yet, for many Pangkhua families, who eventually fled this village, such advantages did not suffice to let them survive. Some of them now live in distant villages in deeper forests where it is easier to cultivate Jum.

In the early 1990s, few families at Mahalu also fled during a crop failure that hit different parts of the CHT where large regions were replanted with bamboo groves to assure raw materials for the paper mills in the Kaptai. In every five to ten years, the bamboos groves starts flowering; the Jumma people traditionally view this as a sign of famine and do not cultivate nearby lands. The flowering causes an oversupply of food for the rodents, enabling them to multiply in numbers within a very short period. When the flowering is over, the rodents become irresistible in the Jum and in most of the cases, people lose all their crops. The Jumma people like the Pangkhuas are also increasingly being subjected to agricultural transformations, mostly a shift towards cash crops like spices and tobacco instead of food grains for self-consumption. A large number of them also switched to fruit cultivation and not surprisingly, failed to profit by selling them in the markets. I saw long queues of boats carrying bananas and pineapples, waiting to be sold at Kaptai, Rangamati, or Bandarban. Ramdin Pangkhua, who used to grow pineapples in Bandarbans before moving to Pangkhupara, told me that once he waited for two days with a boat full of pineapples at Bandarban jetty. The Bengali buyers from Chittagong sat idle for two days watching them anxious about the perishable fruits and finally asked for a price far less than their production cost. Some of them sold it in that price and others, frustrated, threw all their fruits into the Sangu river. That year he could no longer send his children to school as they all had to work in farms. He failed to manage farm land in the following year and had to move here.

As an ethnographer, my experiences with the disjuncture of the Pangkhuas and other Jummas from their land were difficult to articulate with the notions and facts prevalent in the socio-scientific discourses on the same. The first problem for me was that it does not recognise state-making as a fundamental cause of a diverse range of displacements that the people in CHT constantly encounter or vulnerable to. The positivist assumptions and definitions of displacement slip over its actual endurances as I have witnessed at Mahalu or somewhere else. How can we explain such political transformation of places by modern states which extensively use displacement as an essential tool to achieve its political and economic goals in margins like the CHT? Do our spatial forms of knowledge on displacement somewhat play a complimentary role in such transformation by presenting the causal facts as isolated and incoherent? Undeniably such knowledge is necessary to design mitigation measures, but again one may ask does mitigation of isolated cases suffice to end the threats of systematic displacement like the one my friends at Mahalu continuously live with? Does the normative discourse of displacement not undermine the facts about a profound alienation of people from their land evidently engendering from the extraordinary presence the state maintains in this margin? Contemporary anthropology offers several perspectives that can illuminate a political economy of displacement the state enforce to establish itself in such politically punctuated territory. The perspectives also theoretically challenge the spatio-temporal values given to the notion of displacement and help us to map a generic political economy to address the issues of dispossession among the marginal communities.

Displacement as 'Order': The State and Minorities on a Margin

The state is, before anything else, about order; while admitting this fundamental assumption of the political theories concerning state, anthropology has a different point of departure that paves unique ways of rethinking the state and its order making functions. Anthropology, through its vast ethnographic archive, has drawn attention to indigenous groups, inherently defined as non-state peoples presenting a challenge to the modern nation state (Nagengast 1994). As its research hallmark, ethnography is also increasingly gaining voice in the new current of academic scholarship revisiting the 'state' from a liberal, postcolonial or poststructuralist perspectives, whereas traditionally it was not even acknowledged as a proper subject of ethnographic investigation. The presence of the state in classical ethnographies, what Das and Poole (2004: 5) have remarked as 'ghostly', has remained obscured largely due to the colonial construction of anthropology's subject population as 'stateless' societies. Despite the fact that the ethnographic study of so-called stateless societies had always been haunted by the very presence of 'language and figure' of the state, the historical construction of such marginal societies as exception or discontinuation of administrative rationality and political order, has excluded the state as primary objects of ethnographic inquiry. The restoration of the state as proper subject matter of ethnography begins with the assumption that the state is always an incomplete project that requires continual reinforcement, primarily achieved through the 'legitimated monopoly over violence' (Das and Poole 2004: 7). This Weberian notion of the state has been extended by anthropologists arguing that such continuous rearticulation of the state is best evident along its margins and, therefore, the anthropological study of the margins of the state facilitates the understanding of the crucial sites where the state can be experienced in its most critical forms.

Keeping along this line, the reproduction of the CHT as a margin and its people as minorities primarily emerges from a very typical mismatch between nation and state. The political boundaries of the nation state, particularly whose boundaries are actually legacies from the colonial past, often include local bunches of identity and culture incompatible to the nationalist scheme. Most of the nationalist regimes consider such incompatibilities as a threat to the 'holy monument' of their national integrity. Thus, a territorial locale like the CHT becomes a political margin which eventually congregates a historical encounter between power and powerlessness. Eriksen (1991, 1993) views this contrast as a totalitarian versus segmentary ideology in a relationship of exclusion and inclusion. For him, the uniqueness of nationalism lies in a modern, abstract, and 'binary' ideology of exclusion and inclusion, and also in its powerful symbolic as well as practical aspects which he contrasted with 'segmentary' ethnic ideologies (1991: 1). The margin of the state are locales where the appearance of the state is significantly different from the rest of its territory and the experience of the state by the people, who make up, though not exclusively, the minority, is fundamentally different from the rest of the citizenry. The minority refers to the local identity and culture/s incompatible to the nationalist ideology of belongingness. Beside the dimensional differences of the state's presence as order-making functions, the margins of the states are increasingly featuring forms of struggles that challenge the state monopoly over violence (Poole 2004). But this should not be taken as a conclusive perception of the margin or minority. Every margin has its own historical particularity which Lipuma mentions as 'historically constructed character and trajectory' (1997: 61). For instance, small groups like the Pangkhuas in the CHT exhibit a very different 'art of being or not being governed' which simultaneously includes both antagonistic and submissive strategies. The historical transformation from 'clan people' to 'citizens' always involve a political transformation of places where national culture and marginal identity is constantly rearticulated both by the state and locals. From the state's point of view, this articulation is 'given and accepted as progress' and does not recognise prevalent indigenous identities beyond itself. The limits of the nation state, its region and its cultures 'exist only as they are imagined, institutionalized and contested in the public sphere' (Lipuma 1997: 62). The culture of nationhood, what Lipuma suggests as 'the culture of empowered consensus', is also a form of such contestations.

What, in the context of CHT, 'order' means for the Bangladeshi state is largely defined by few of such 'empowered consensuses' and 'historical trajectories' embedded in the nationalistic claim it puts on the CHT. For example, Since the 1960s, perhaps everyone has the same answer to what is the first economic and development challenge for the Bangladeshi state – its overpopulation. The implicit public view of the CHT was built upon the fact that this 10 per cent of the national territory was inhabited by less than 5 per cent of the population until the 1960s. This provided justification to the resettlement of the Bengalis in the region which pushed the density figures almost equal to the national average by the 1990s. The resettlement programme along with the environmental impacts of the Kaptai lake, fundamentally reconfigured people's relationships with the place, for both settlers and indigenous people. The way the state redefines human relationship with the territory depends on its economic and political perception of the given territory. In the case of the CHT, the Bangladesh state possesses an exclusive economic imagery of the region and uninterrupted resource extraction becomes the main concern of what it enforces as 'order'. Most of the infrastructural developments in the CHT are clearly dedicated to facilitate transportation of forest resources and security arrangements are also focused on regulating these trades.

However, the order-making role of the state in the CHT also has a transnational dimension. The mobility pattern in the CHT is historically linked to a greater flow that emerged from the colonialisation of India's northwest and to some extent, Burma. Unlike the rest of British India, the colonial state penetrated here more in terms of economic exploitation than establishing political hegemony. Such colonial expansions later formed the basis for an ethnic and nationalist hegemony by the Bengalis that was ossified and challenged in postcolonial South Asia. Historically, the Bengalis, one of the central ethnic traditions of the Indo-Aryan civilisation, also were the native forerunners of colonial modernity. The colonial investments in booming economic sectors like timbers, cotton, or tea plantation extensively employed Bengalis across Assam, Meghalaya, Tripura, and though distinctly, the CHT. This led to a major demographic shift for the Bengalis who are still engaged in political conflicts in places of what many call the 'Mongoloid Fringe' of India (Bhaumik 2005: 144). Like the case of the CHT, the Bengalis in Assam and Tripura also form a dominant political force of elite nationalist bourgeoisie that runs all the state governments in South Asia. Still, the elite nationalist politics is not without inner contradictions. The nationalist politics in Bangladesh also has two feuding camps who swapped power in each of last four parliamentary elections. The first, led by the Bangladesh Awami League endorses 'Bengali' nationalism, which is presumably inclined to a secular imagination of nationality. The other, which consisted of the Bangladesh Nationalist Party and its Islamist allies, proclaims to be 'Bangladeshi' to demarcate differences from rest of the non-Muslim Bengalis living in West Bengal or other parts of India. Both camps also blame each other for being 'pro-Indian' and 'anti-Islam' or 'anti-Indian' and 'communal'. The rivalries and its regional linkages always had significant resonances over the way the state maintain order in the CHT. The CHT has been reported for cross-border insurgency, terrorism, arms, and drugs smuggling. In many instances, the Indian and Bangladeshi governments have accused each other for harbouring insurgents across the CHT border (Mohsin 1997: 207). Although, the Bangladeshi state has a complex and altered role in relation to these cross-border issues, these were constantly used to legitimise a permanent 'emergency' rule in the CHT.

So, as an 'emergency zone', the order-making practices in the CHT is characterised by a legitimised 'exception' of the normative rights of its citizens which means that the state is unable or unwilling to fulfil its own legislative mandates here. Bangladesh claims to have little control in this region and, therefore, insists on a violent surveillance mechanism and arbitrary controls over mobility, settlements, and, nevertheless, life. The stately use of displacement as an order-making tool also enacts specific economic and political goals which relegate people's own rights, efforts, and ways to form locality. Thus, the resettlement of the Bengalis, construction of Kaptai dam, ethnic conflicts, counter-insurgency operations, alienation of indigenous groups from subsistence economy, deforestation, plantation, industries or the military bases become parts of an inter-linked mode of economic and political dominance which the state perceive as 'order' in the CHT.

References

Ahmed, I. 1996. *State, Nation and Ethnicity in Contemporary South Asia*. London: Pinter.

Ahmed, R. 1991. *Krantikale parbattaya chattagramer jonogoshthi: Ekti porjalochona* (The people of CHT in transition: An overview). Nribiggyan Potrika, 2, 58–57.

Amnesty International, 2000. '*Bangladesh: Human Rights in the Chittagong Hill Tracts*'. Report ASA 13/001/2000, <http://www.amnesty.org/en/library/info/ASA13/001/2000/en>. 1 February 2000, (Accessed 8 September 2007).

Barua, B.P. 2001. *Ethnicity and National Integration in Bangladesh: A Study of the Chittagong Hill Tracts*. New Delhi: Har-Anand Publications Ltd.

Bhaumik, S. 2005. 'India's northeast: Nobody's people in no-man's-land', in *Internal Displacement in South Asia: The Relevance of the UN's Guiding Principles*, edited by P. Banerjee, S.B.R. Chaudhury, and S.K. Das. New Delhi: Sage Publications, 144–74.

Das, V. and D. Poole (eds). 2004. State and its margins: Comparative ethnographies, in *Anthropology in the Margins of the State*. New Delhi: Oxford University Press, 3–33.

Dewan, H., S. Chakma, H. M. Chakma and P. Chakma (eds). 2006. *Ranya Phool: Selected Changma Poetry*. Dhaka: Mani Shwapan Dewan.

Eriksen, T.H. 1991. Ethnicity versus nationalism. *Journal of Peace Research*, 28(3), 263–78.

Eriksen, T.H. 1993. *Ethnicity and Nationalism: Anthropological Perspectives*. London: Pluto Press.

Guinness, P. 2002. *Migration*. London: Hodder Education.

Kabir, M.H. 1998. The problems of tribal separatism and constitutional reform in Bangladesh, in *Ethnicity and Constitutional Reform in South Asia*, edited by Iftekharuzzaman. Dhaka: The University Press Limited, 10–26.

Karmakar, A. and H. K. Chakma, '*Tiranobboi Hajar Pahari Poribar Ekhono Udvastu*'(Ninety three thousand Hill Families are still Refugees), Daily Prothom Alo, 19 August 2009, 18.

Lipuma, E. 1997. The formation of nation-states and national cultures in Oceania, in *Nation Making: Emergent Identities in Postcolonial Melanesia*, edited by R.J. Foster. Ann Arbor: The University of Michigan Press, 33–68.

Mey, Wolfgang. 1996 [1980]. *Parbatya Chattagramer Kaumasamaj: Ekti Artha-Samajik Itihas* (Politische Systeme in den Chittagong Hill Tracts, Bangladesh), trans. Swapna Bhattacharya (Chakraborti). Calcutta: International Center for Bengal Studies (ICBS).

Mohsin, A. 1997. *The Chittagong Hill Tracts and the Politics of Nationalism*. Dhaka: The University Press Ltd.

Mohsin, A. and I. Ahmed. 1996. Modernity, alienation and the environment: The experience of hill people. *Journal of the Asiatic Society of Bangladesh*, 41(2), 265–86.

Mooney, E. 2005. The concept of internal displacement and the case for internally displaced persons as a category of concerns. *Refugee Survey Quarterly*, 24(3), 9–26.

Nagengast, C. 1994. Violence, terror, and the crisis of the state. *Annual Review of Anthropology*, 23, 109–36.

Norwegian Refugee Council. 2002. *Internally Displaced People: a global Survey*, London: Earthscan Publications Ltd.

Poole, D. 2004. Between threat and guarantee: Justice and community in the margins of the Peruvian state, in *Anthropology in the Margins of the State*, edited by V. Das and D. Poole. New Delhi: Oxford University Press, 35–66.

Rahim, A. 1997. *Politics and National Formation in Bangladesh*. Dhaka: The University Press Ltd.

Scott, James C. 2009. *The Art of Not Being Governed: An Anarchist History of the Upland Southeast Asia*. New Haven: Yale University Press.

UNHCR, United Nations Commission on Human Rights. 1992. '*Analytical Report of the Secretary-General on Internally Displaced Persons*'. UN Doc. E/CN.4/1992/23. 14 February 1992, Para. 17. http://daccess-dds ny.un.org/doc/UNDOC/GEN/G92/106/61/PDF/G9210661.pdf?OpenElement (accessed 18 August 2009).

van Schendel, W. 1992a. The invention of the 'Jummas': State formation and ethnicity in Southeastern Bangladesh. *Modern Asian Societies* [Online], 26(1), 95–128. Available at: http://www.jstor.org/stable/312719 [accessed: 8 September 2009].

van Schendel, W (ed.). 1992b. *Francis Buchanan in Southeast Bengal 1798: His Journey to Chittagong, the Chittagong Hill Tracts, Noakhali, and Comilla*. Dhaka: University Press Ltd.

van Schendel, W. 2005. *The Bengal Borderland: Beyond State and Nation in South Asia*. London: Anthem Press.

PART 3
Placeless Identities: Non-State Places and Floating Peoples

Chapter 10
Home-making and Regrounding: Lives of Bangladeshi Migrants on the Damodar Charlands of Lower Bengal, India

Kuntala Lahiri-Dutt and Gopa Samanta

Introducing the Field

This chapter focuses on the experiences of home-making and regrounding of mobile subjects of women and men on the uniquely mobile environment of the charlands in lower Bengal. To survive in this shifting, volatile nature gives rise to a sense of place, which comprises the physical geographical context shaped by the social and cultural capacities of adjustment and the politics of illegality that stigmatise the transborder migrants living on the chars. In this chapter, we explore the Livelihood experience of women living on chars located in a very small area of the lower Damodar valley in the Indian part of the Bengal Delta. We show how women here play a critical role in making a living from minimal resources and develop intimate knowledge of and connections to the land, water, and resources, as well as develop new skills to cope with adversities on a day-to-day basis.

The char dwellers cope with the river's dispositions, described here as 'dancing with the rivers' – to be able to know and constantly adjust to the changing moods of the river. We note that while this coping is strategic, it is also purely contingent and temporary, and, hence, should not be conflated with ecological terms such as 'adaptation' because the residents do not perceive these chars as a permanent 'home'. The study brings forth the multiple sites of 'home' and multiple displacements by exploring women's notions of physical dislocation, banishment, and exile, and also the continuations in particular of cultural traditions, in regrounding themselves in an alien environment. It illuminates the importance of place – not just as a thing in the world, but as a way of seeing, knowing, and understanding it – and men and women's different experiences of home-making. The feminine experiences of the habitus in rebuilding the homes post-displacement are underlined in this chapter. The experience of place, however, is not universal and gender-neutral, and men and women on the chars do not form similar relationships to this unique environment. We follow the feminist geographer Gilian Rose (1993: 53) who posed a powerful critique of Tuan's (1977, 1974) conceptualisations of making places and home-making as gender-neutral. She says: 'This enthusiasm for home and for what is associated with the domestic, in the context of the erasure of women from humanistic studies, suggests to me that humanistic geographers are working with a masculinist notion of home/place.' Communities can be stifling and homes can be, and often are, places of drudgery, abuse, and neglect, Rose continues (1993: 55), and the narratives of Bangladeshi women living on the chars show how the home is made with their 'blood and tears'.

Angelika Brammer (1994: xi) has offered a succinct definition of displacement as an analytical construct: '[D]isplacement refers to the separation of people from their native culture, through physical dislocation (as refugees, immigrants, migrants, exiles or expatriates) or the colonizing imposition of a foreign culture.' This clearly brings out the unity of the personal experiences of displacement, highlighting the futility of classifactory approaches in understanding displacement. Anderson and Lee (2005: 11) have shown that lived experiences of categories such as the immigrant, refugee, expatriate, and migrant, and physical/spatial displacement, cultural displacement, psychological/affective displacement, and intellectual displacement are not mutually exclusive. Physical dislocation for any reason or the displacement of ancestral culture may transform human connections to place and dislocate the sense of place. In this chapter, we will show that the sense of place and placelessness are also intricately connected in the perceptions of the Bangladeshi Hindus who come to live on the charlands.

Although women are 'dependent migrants', having followed the male members of the family from Bangladesh, they play a major role in providing livelihood support to sustain their families. We add a geographical perspective to the increasing understanding of gendered experiences of displacement by understanding their coping processes to the environmental dynamics of the charlands. The participants in our study, mostly Hindus, had moved illegally from Bangladesh to India along with their families. We show the complex interplay between poverty and insecurity of life and property through a gender lens to identify the pattern of uprooting and reorganisation of livelihoods on the chars. For some, poverty has been a factor and cause of migration. Some families were physically forced to leave Bangladesh, leaving behind properties (houses, agricultural land, business shops), whilst others left because of a feeling of insecurity from the rising Muslim fundamentalism. For some, displacement has led to perceived poverty, painful experiences, and difficulties in cultural adjustments during the early phase of settling down. In general, for women living on charlands, displacement was not only a painful experience, but also involved heavy work, isolation from the older support networks that sustained their social connections, and, hence, both the cause and result of poverty.

Accessing Chars and Charland Dwellers

This chapter emerged from a broader research project on hybrid landscapes and the livelihoods of people on the charlands of deltaic Bengal. Intensive field-based empirical research was carried out in different phases from 2002 to 2010 for this self-funded project. The study area constitutes a small part of the Lower Damodar Valley, within Burdwan district of West Bengal in eastern India (see Figure 10.1) where both of the authors undertook doctoral studies in the 1980s and 1990s on agrarian change and urbanisation. The chars that were studied in this field research were the ones that were more accessible from the northern side of the river embankment. The social worlds of the chars were entirely different to us, even though we had carried out research in the area previously and we lived just across the embankments. The illegal status of charuars also meant that we needed to exercise caution and ethical judgment. As outsiders, securing access to the charland people's homes and lives was not a straightforward and easy process. During the first year or two, we spent a day almost every week casually walking along the different paths and tracks on the chars, and made small talk with those who had the time to spend. These casual conversations built the familiarity and trust to allow entry into the community life. It also enabled us to observe people's everyday lives in detail, observations that were useful when we formulated the questionnaire for a later survey during the second phase. In this phase, we carried out a complete household census based on a structured questionnaire of all the 1,312 households living in a total of eleven chars.

During the third and fourth phases, we spent more time trying to understand the migration process, the complexities of survival in daily lives, money management and the experiences of home-making. These studies were largely qualitative in nature. Following Limb and Dwyer (2001), we relied on the understandings and knowledge of the char-dwellers gained through interviews, discussions and participant observation. Many of the 80 women and 150 men whom we interviewed have remained in touch with us. These men and women were selected arbitrarily – from different age groups and economic classes as a matter of convenience, without following a strict statistical sampling method.

This paper uses material from in-depth interviews with 20 men and 15 women; these parts focussed on migration and post-migration livelihoods. Although arbitrarily chosen with no conscious effort to undertake 'scientific sampling'; the resultant interviews can be described as a 'sample' of lives of charland men and women who willingly participated (as envisaged by experts on ethnographic research such as Hammersley and Atkinson 1995). At this stage richer ethnographic data emerged, that allowed us to build a strong understanding of the entire dynamics of displacement from Bangladesh, experiences of migration and regrounding on these chars, and the perception of 'home'. No specific set of questions or a key question was used as we only tried to participate and observe their everyday lives and gave a patient hearing to the stories as individuals recollected. In listening to their stories and their experiences, we often interrupted the participants for facts and specific queries. We refrained from taking notes in front of them to avoid missing any part of the conversation and to avoid suspicion, and wrote the notes after visiting the field. Besides individual interviews, we also had group discussions with six or seven participants. These too were opportunistic; we would sit on someone's open veranda to have an informal chat, and as passers-by strolled along, they would choose to join in depending on their interest in the topic. Gradually a motley group would gather around us.

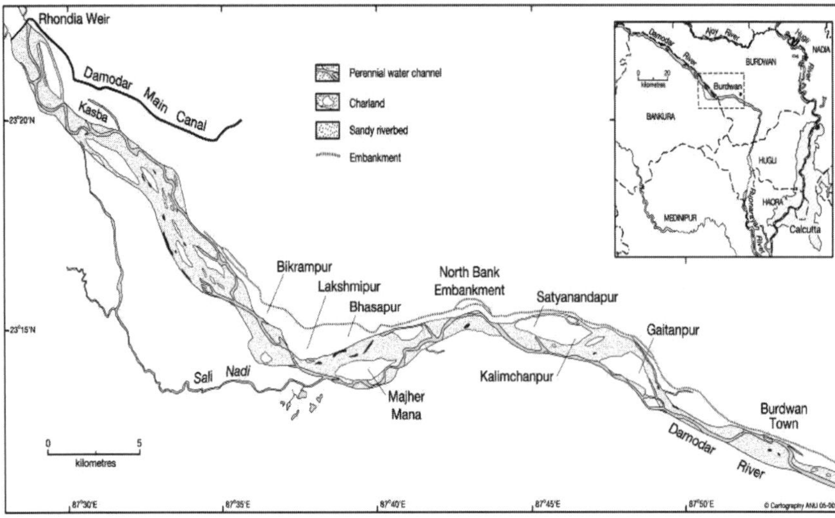

Figure 10.1 The Chars under Study

Source: Fieldwork data and Survey of India topographical sheets. Cartography by Kay Dancey, ANU.

Making Sense of Charlands and Their People

How Chars are Formed

The Bengali term char means a piece of land formed by fluvial deposits on riverbeds. There is no dearth of charlands in Bengal. Some of the mighty rivers arising in the Himalayas carry huge amounts of silt and descend onto the Gangetic plains to flow only very sluggishly on the near-perfect flat plains for most of the year, only to metamorphose into devastating torrents during the monsoon to inundate the plains. The roaring rivers almost choke with the enormous body of sand and other sediments carried in their waters. The sandy alluvium creates uninhabitable, uncultivable diaras in north Bihar and eastern Uttar Pradesh flats, and more expansive and more flat chars further down the plain in Bengal Delta. Within a few years, the humid climate assists the growth of grass and reeds, and a slow process of organic breakdown facilitates refertilisation of the lands. From a geomorphological point of view, chars are described as 'sand-bars' or attached bars that are formed near river banks and 'river islands' when they form on the riverbeds (in 'braided channels').

Whilst sediment accretion gives rise to charlands, they are also constantly eroded by the rivers that eat up their banks and shift their courses frequently. Majuli, the largest char in the world formed by the great river Brahmaputra in Assam in northeast India, has existed for at least two-and-a-half centuries. Newly formed chars are naturally transitory and fragile; one may even disappear overnight. Bank erosion is as much part of char dynamics as accretion; Majuli too suffers from a high rate of bank erosion. Chars are diverse in nature: the specific pattern of their physical development and human use of land and other resources may differ from country to country, one river system to another and even vary within different reaches of the same river. Generally the chars are surrounded by water, but heavy flows may even completely inundate low-lying chars. In many seasonal rivers with braided channels, some island chars may only be accessed by crossing a river channel even in the dry season.

Most newly emerging chars, being on the borderlines of land and water, do not exist as legitimate and officially recognised pieces of land. Lying outside or at the margins of the land revenue system, chars represent a hybrid and fluid landscape that provides opportunities to some people who lives literally 'on the edge'. Most chars may house permanent human settlements, depending upon the silt and soil properties and the strength of river currents. A recent UNDP report notes that there are over 3 million people living on chars in Bangladesh. No comparative estimate exists for the chars in India.

Chars (and Choruas) in Bengal

Our research follows Abdul Baqee's 1998 research on the occupancy process of charlands in Bangladesh. Baqee calls chars as 'Allah jaane' lands ('God only knows', or 'whatever will be, will be'). Following Shapan Adnan's 1976 account of stark rural life in Bangladesh, and Currey (1986) and Ali's (1980) meticulous research, Baqee focuses on the process of occupancy, dislocation, and resettlement. For example, he describes the violent process of occupation over newly rising chars. In 1979, in Char Kashim of Chandpur in Bangladesh, about 300 families who were living with an annual lease from the government were attacked by people wearing police uniforms, and those who resisted were abducted by powerful local leaders who hired lathiyals (stick-wielding local force). At the same time, he points out that the laws governing charlands of Bangladesh date back to a time when there was less population pressure on land. According to him, the covert understanding between local matbars (leaders) and corrupt state officials at the local level, and faulty land survey that is ill at ease with the changing nature of charlands give rise to violent and complicated disputes often ending in bloodshed. Later research on charlands in Bangladesh has been at the centre of resource management and environmental policy debates, discussions around flood mitigation, and human vulnerabilities caused by river-bank erosion, and development-funding by foreign agencies. Chowdhury (2001) considers that the greater awareness of chars as populated places in Bangladesh is understandable because of the importance of the riparian areas to the country's life and economy. She also touches on gender aspects in her study of disaster mitigation in charlands in Bangladesh. Comparatively, almost no research has explored social dynamics of charlands in West Bengal and elsewhere in India, and charlands have remained virtually isolated from the 'mainstream' literature on mobility and displacement, gender and livelihoods in resource constrained contexts.

The Fluid Landscapes of Damodar Chars

Charlands located in the lower Damodar valley in West Bengal are different from Bangladeshi charlands in a number of ways; geomorphologically as well as from the economic point of view. These chars are formed of coarser sand than those in Bangladesh. Lying within the administrative boundaries of Burdwan and Bankura districts, and located on the western edge of the Bengal Delta, on the western bank of the Bhagirathi-Hooghly branch of the Ganges River, the eastern part of Burdwan that has been traditionally known for rural prosperity, they are also close to one of the richest farming areas of this region. Part of the rural prosperity came from trade between lower parts of the delta on the east and forest-covered undulating tracts of the Chotanagpur Plateau in the west. The trade was, until early nineteenth century, largely river-borne, and Damodar carried large vessels up to the mid-nineteenth century.

Damodar has been one of the more notable rivers of Bengal also because of its role in flooding, for frequently changing its course, and for causing drainage congestion and bringing prosperity to those living in its valley. The Damodar Valley Corporation (DVC), formed in 1948, was meant to administer multipurpose development in the basin through a series of dams, barrages, and canal network. It has been only partly successful in reducing the frequency and intensity of floods and in providing irrigation water to agricultural fields of the Lower Damodar Valley. Saha (2008) observes that the dams and barrages have led to remarkable changes in river ecology and physical environment in the lower reaches of the valley. Moreover, Bhattacharyya (1998) has shown that dams on the river have reduced downstream water flow and altered the river regime adequately to give rise to chars. According to her, river control was primarily responsible for the formation of chars in the lower reaches of Damodar. She shows that the sluices enhanced the formation of innumerable chars on the riverbed by releasing coarser and heavier sand through lower parts of the gates. Elsewhere, we have shown that the meanings of land and river to those who live on or near them and those who govern them have been historically responsible for the treatment the charlands received (Lahiri-Dutt and Samanta 2007). These meanings have evolved over a complex process of pre-colonial, colonial, and post-colonial social, legal, and political processes from which neither land nor rivers of Bengal remained isolated. What we could ascertain from interviews and anecdotal evidences was connections between the decreasing frequency and changing nature of floods with the construction of dams and barrages on Damodar, and the growing number of human settlements on the chars.

Continuous immigration from Bangladesh has increased the population significantly in these areas. In most cases, agricultural lands adjacent to a house are usually 'owned' by the householder. This arrangement has facilitated the employment of family labour, especially women, in the field. The soil, though only a few inches thick, is fertile for the production of several crops. Sand fields are gradually converted into agricultural croplands by bio-manuring and hard labour of the migrants. During our field survey, we observed almost everywhere sand being reclaimed of for crop production.

High water requirement of these sandy soils means that only those with shallow and submersible pumps can tap groundwater to produce paddy. Poorer farmers keep their lands fallow during the kharif or monsoonal main cropping season (July to October). The winter season, starting from October and on till March, is the main cropping season. Relatively better-off farmers produce paddy for both seasons of the year (kharif and boro, or winter crop). The farmers producing vegetables engage women of their families to provide the extra attention that these crops need. Poorer women often work on others' fields when their own lands are fallow during the kharif season. The fallow during kharif provides a good opportunity for animal rearing especially of cows, goats, and pigs that supplement family incomes. This time of the year is rainy, giving rise to a good growth of local grasses for the animals. The temporary grazing fields and activities are also supervised by women, adding an extra burden to the already busy lives.

Houses in char settlements are usually aligned in a linear fashion on the two sides of a dirt track that runs through the middle of the land. The peripheral areas are used as agricultural land. Built of bamboo (the dominant plant species in the char) and mud, with paddy straw or corrugated tin and asbestos roofs, the houses are as temporary as their occupants. Poorer families' homes have only bamboo walls, that are unable to keep the chill brought in by winter breeze brushing over the sand. Uma Mondal, says: 'The nights in winter in our house are so terribly cold that they make me look forward to the morning, even though the day is full of work for me.' Recently, a handful of residents – mainly those who had arrived after the devastating flood of Damodar in 1978 and have yet to experience the fury of the floods – have built pucca (concrete) houses on the riverbed with loans from local mahajans (moneylenders). The idea was that a bigger investment would fetch better prices if and when they leave the char in future.

Cross-border Mobility of the Bangladeshis

Anderson and O'Dowd (1999: 594) view borders as interfaces between competing territorial political units, and, therefore, they are deeply embedded within these systems. However, Williams and Baláž (2002: 651) have shown in their study in Eastern Europe that when borders are porous, they can act as gateways to economic opportunities rather than as barriers to economic linkages. The partition in 1947 saw the movement of about 13 million people across the border; this is claimed as one of the largest inter-country transfer of population in the world (Haque 1995: 189). There were major differences in population displacements in West and East Pakistan; in the eastern part, many Bengali Hindus did not initially want to leave their homes. Soon, people were being pushed out of their villages and those crossing the border to enter India illegally were uprooted (Rahman and Schendel 2003). Unlike the divided Punjab, population migration across the border in divided Bengal was carried out slowly. Also, in the initial years, the refugee influx, or at least its historical documentation, was primarily focused on the Bengali bhadralok (educated upper and middle classes). As communal violence and riots escalated in post-partition years, many more Hindus began to leave the then East Pakistan for India after 1947 (Chatterji 1995). In the following decades, about 10 million Bengalis took shelter in the neighbouring Indian states of West Bengal and Assam due to political and religious suppression (Alam 2003). These undocumented, unsung migrants – bastutyagis (those who left their homes), bastuharas (those who lost their homes), and sharonarthis (seeking refuge) – from the interior are literally the 'midnight's orphans', who decided to (or had to) remain in East Pakistan after Partition, but found themselves increasingly marginalised as the newly founded nation of Bangladesh decided to adopt Islam as 'state religion' and religious fundamentalism gained momentum among the poor peasants in the country (see Human Rights Feature 2001; Mohsin 1997). The Hindus and other religious minorities in Bangladesh have responded by adopting migration as a means of escaping the persecution and to secure a livelihood. The 1990s experienced a hardening of allegiances that conflate religion and nation in South Asia. Rising religious fundamentalism within India, and particularly the rise of Bharatiya Janata Party (BJP) at the helm of power in New Delhi hardened the stance towards the minorities in both India and Bangladesh. The aftermath of 2002 elections in Bangladesh was particularly violent; Datta (2004: 337) estimates that about 5,000 to 20,000 Bangladeshi Hindus and other minorities fled to India to escape violence. With each act against the Muslim community in India, Hindus in rural Bangladesh faced retaliatory attacks. Many of them eventually left with their families and dispersed within Indian populations (Bose 2000).

Not only Hindus, but a steady stream of Bangladeshi Muslims has also migrated into India due to the sheer weight of poverty. For a small deltaic country, with the eighth largest population in the world, where natural disasters and political uncertainties are regular features, Bangladeshi populations have to seek refuge in migration (Costello and Mahendra 2001; Siddiqui 2003) – both legal and illegal migration (Samad 2004). Seen as a flight from poverty, migration offers little opportunities to the migrant in the new land and, thus, generating both the cause and result by poverty (Skeldon 2002). The choice of destination and levels of benefits and risks vary significantly according to economic and social power.

The complex context of environmental factors, excessive population pressure on productive land, political instability, extreme poverty, and rampant corruption force Bangladeshis to migrate more than other nationalities. Samaddar (1999) accepts the Bangladeshis as comprising an 'immigrant niche' in general, but contests the 'population pressure leading to migration' thesis. He shows that migration from Bangladesh to West Bengal is rooted in complex reasons – environmental hazards such as floods, droughts, tropical cyclones and tidal surges, as well as religious, political, ethnic, economic, and environmental reasons.

Political instability also acts as a major contributing factor to migration. Increased radicalisation of Islamic rural communities and the rise of religious groups in recent years in Bangladesh have created a turbulent socio-political milieu, leaving the minority communities feeling insecure. Persecution of the religious minorities, primarily the Hindus, the Christians, and a few 'tribal' groups such as the Chakmas who live in Chittagong Hill Tracts, has also surfaced in international media. The most sensitive amongst these is the migration of the Bangladeshi Hindus to India to escape religious persecution. At less than 6 per cent of the population, the Hindus are still the largest religious minority in Bangladesh. While other minorities tend to bear the cross silently, the Hindus in Bangladesh respond to this trouble by leaving the country (Samad 2004). A number of authors, for example, Barkat and Zaman (2000), Malik (2000), and Samad (1998) have documented the forcible eviction from landed property of the Bangladeshi Hindus. In rural Bangladesh, the targets have generally been men and women from poorer and more disadvantaged groups. In addition, two discriminatory laws – the Enemy Property (Continuance and Emergency Provisions) Act 1974 and the Vested and Non-Resident Property (Administration) Act 1974–contributed significantly to driving Hindus out of their lands in Bangladesh and ultimately pushing them beyond the border to India.

Silent Footfalls

Cross-border migration has lately been at the heart of international political debates. One strand of this debate views these 'outsiders' posing a threat to national security, triggering fear that they would, in the long run, 'take over' the economy (at least in the form of cheap labour), society, and culture of the 'insiders', even though today's insiders were all originally from within the modern state concerned. The question of illegal migration is complex also because of the lack of accurate numbers; estimates of illegal migrants are often based on indirect observations. Unfortunately, there is no comparable data for India; only some guesstimates are available. The Border Police Department estimates that about one thousand cross border each day and enter West Bengal (Mukherjee 2003). Datta (2004) of the Indian Statistical Institute, Kolkata, estimates that 15 million Bangladeshi nationals are living in India illegally. Another report by Pathania (2003) suggests this figure to over 5 million only in West Bengal. So sensitive is migration research in eastern India as a subject that battle lines are clearly drawn, and objectivity is difficult to achieve for a researcher. Populist views claim that the very idea of West Bengal, created as a non-Muslim majority state, is at stake due to the mass influx of the Bangladeshi Muslims into West Bengal (Ray 2009), and that on a number of occasions (such as in 1964, 1990, 1992, and most recently, in 2002) pogroms have deliberately targeted the Hindus in rural Bangladesh (Roy 2001).

Transborder movements without legal documents – 'Illegal' or 'unauthorised' migration – is not uncommon within Asian countries (Battistella and Asis Maruja 2003: 11–13). In South Asia in particular, political boundaries that were established during the colonial time are often not much more than a line on the ground and are surrounded by disputes. Consequently, migration from one country to another are issues fraught with controversy and expose raw edges of sentiment (Chakraborty et al. 1997). Media hypes about 'infiltration' have been critiqued in the media itself (see Bandyopadhyay 2004, for example). Politically, India–Bangladesh relations have been the subject of some stress ever since the latter achieved independence, and whether it is the sharing of river waters or Farakka Barrage Project on the Ganga River or Tin Bigha or New Moore Island, the two countries have not given up any opportunity to cross swords (Ahmad 1997; Jacques 2001).

Two key authoritative pieces of research have posed a critique of the conceptualisation of borders as absolute and key political instruments of statecraft and nation-building projects in South Asia. Both agree that border formation was a colonial project; subsequently these borders as boundary lines were internalised by the post-colonial states. Samaddar's 1999 work on transborder migration from Bangladesh shows how such borders have become the markers of statehood in South Asia. Yet, he argued, these borders are unstable, and produce even more partitions by marking out the separatedness of the various solidarities. As the alien becomes a separate and oppositional subject, both nation and citizenship become utilitarian categories. Questions of illegality of transborder flows then get conflated with those of national security to acquire politically explosive dimensions. More recently, Banerjee (2010) differentiates between frontiers and borders, and talks of the pace with which the frontiers were turned into borderlands with little knowledge of what it actually entailed. As frontiers hardened into political borders, those at the 'core' denied the existence of real people along, across, and around these borders, people who inhabit, who cross, and try to appropriate the borders themselves. Banerjee's (2010: xxxiv) point, that 'it was the partition which changed them [the frontiers] into borders' is most relevant in understanding the cross-border mobility of the Bangladeshi Hindus into the char (and other) lands. Such mobility challenges the prevalent systems of power and theories of sovereignty that assumes these borders as static lines. The presence of the Bangladeshi Hindus on Damodar charlands is proof that the borders are alive and have different meanings and forms. Our research participants raised the question of legality and illegality of tranborder movement in their own ways, and compelled us to think about the transborder movement from their perspective. Biswanath Mondal says:

> Tell me, how can I become an illegal migrant to India? I was born in the Bangladesh part of undivided India, but I grew up on this side [of the border] in West Bengal with my uncle. I went to college in Katwa [a town in Burdwan

> District]. Suddenly, on 14 August 1947, somebody told me that I am not an Indian! I came to be known as a Bangladeshi because my family and my land and property were on the other side [of the border] in Bangaladesh. When I later came back here with my family I became known as an illegal migrant to India. Believe me, I do not see myself as an illegal person. I should also have the same right to this land as you.

Such statements defy the very notion of fixed citizenship that is proposed by neoliberal states. based on ideas of fixed citizenship. These are notions that arouse fear and sense of insecurity; Samaddar (1999: 17) observes that transborder mobility has become a sensitive political issue: '[T]he spectre of hordes of illegal immigrants … constantly haunt the pundits and practitioners of statecraft in South Asia.' The question of ill-treatment of the minorities gets overshadowed by the degradation of the environment which also leads to displacement of a large number of the poor who are thought to flock to India as environmental refugees. Since the rule of BJP in India in 2003, the topic of cross-border displacement has become politically 'hot', as the then Deputy Prime Minister Lal Krishna Advani advocated an uncompromising approach towards undocumented Bangladeshis, by issuing a national directive to all provinces to take 'immediate steps … to identify them, locate them, and throw them out' (Ramachandran 2005). The Hindus in Bangladesh have been described as 'stateless' and 'the marginal men' in the post-partition Indian subcontinent (Chakrabarti 1990). Such migrants thus constitute a 'marginal nation' labelled as 'nowhere people' – without valid papers on the no-man's land between two countries (Acharya et al. 2003).

Scholars have contested this scenario and have stated that the border not only fixes the notion of illegality to such migration, but also plays a crucial part in determining the responses by the state. As Samaddar (1999: 21) states, questions of 'illegal' immigration, combined with those of national security, have acquired politically explosive dimensions and have created a marginal nation of transborder displaced who cross the artificial political borders to maintain historical and social affinities, geographical contiguity, and economic imperative. In South Asia, therefore, cross-border migration disrupts the conventional, spatial imagery of nation states and signals a deterritorialisation of communities. Whilst we follow Samaddar's vein of thinking, we also show how new communities are formed as new homes are made in new environments. We also explore the significance that gender adds to the problematic of cross-border displacement and show that post-migration responsibility falls mostly on women, who also face increased burdens in the absence of familiar contacts established back home. The literature on Bangladeshi migration to India is commonly silent on the question of gender, taking displaced communities as universal excepting in partition studies (Ramachandran 2003).

Making a Home on Charlands

Social and economic status of lower caste Bangladeshis after migration to India was rather low. Mallick (1999: 104–5) describes those evicted and forced to go back to Marichjhapi as 'untouchable refugees', defined by Jalais (2005: 1758) as nimnabarno or nimnabarger lok, such as those from the lowest or the Shudra caste. The nimnabarno communities (lower-castes) became politically marginalised minorities in both countries. During or right after the partition, it was the upper-class, upper-caste, better-off elite Hindus who left East Pakistan, leaving behind the rural, poorer, farming castes. These groups were shuttled around in unlivable marshy wetlands or sent away to the remote and undulating forests-clad Dandakaranya resettlement colonies which were surrounded by hostile local tribals. The cruel treatment provided an added impetus to sneak into India illegally. Some obtained citizenship documentation through local officials or political patrons after settling in no-man's land like the chars, whereas others tried to acquire such papers after arriving in the chars.

Although the settlers of Char Gaitanpur crossed the border illegally, only a handful came directly from Bangladesh. Initially, border districts like Nadia and 24 Parganas (both North and South), provide temporary stays; changing residence three or four times thenceforth is a common practice. Often, the migrant visits India several times using family or other connections to go back to Bangladesh frequently, before setting finally. Migration pathways in the chars are complex. Their movements are complex systems of circulation between two or more destinations, through a network rather than as individual cases. An integrated network of family, kin and community in these chars has provided access to most migrants. Migrants coming directly often had strong kin relations who were already settled, and willing and able to arrange work and shelter for the new arrivals. In most cases migrants left home after establishing these connections.

Migrants in the char are mainly from the Khulna and Jessore districts of Bangladesh, which are proximate to the border of West Bengal. They had not previously lived on chars and had no experience of coping with the river's changing moods. Living in the char environment, therefore, meant a new beginning for them. Ratan Mondal ascribes this adaptation to 'adjustment' borne out of sheer need to survive – 'Banchar tagid' as he describes it. He equates himself and other char dwellers with worms living on the soil: 'Amra matir keeter mato, mati kheye benche thaki.' (we are like earthworms, feeding on the soil). They, however, moved into the chars with their farming skills honed from generations of tilling the muddy clay of Bengal, and cleared up the grasses, weeds and reeds to turn the uninhabited lands into fertile agricultural lands. The migrants crossed the border of North and South 24 Parganas and started to live in Indian towns such as Basirhat, Hijalganj, Chaital, Itinda, and Habra – the main entry points for the migrants. In only three cases, the families went back to Bangladesh because they could not arrange to get citizenship papers, or because they could not find work or shelter. Most migrants came to the chars after spending considerable time – six months to five years

– in different locations elsewhere, often closer to the border. They chose these charlands because of plenty cultivable land, not easily available in border districts.

Social networks are an important source of social capital and play an immensely important role in the migration of the char dwellers. Some immigrants, such as Prasanta Mondal (locally known as 'doctor' as he performs the role of a 'quack' besides his agricultural activities) in Char Gaitanpur and Mahadeb Sarkar of Char Kasba, played key roles in the charland migration network. These men were early settlers on individual chars, and served as contact persons who arranged livelihoods for their relatives and village men. This might have given them a sense of power in rebuilding the social networks they had left behind. The influence of kin and neighbour networks during the occupancy process resulted in concentrations of Bangladeshi migrants from different districts in different charlands. For example, in Char Gaitanpur, the migrants are from the Khulna District of Bangladesh; in Char Kasba, they are mostly from the Gopalganj subdivision of Faridpur District; in Char Majher Mana, they are from Barishal; in Char Kalimohanpur, they are from Jessore; and in Chars Lakshmipur, Bhasapur, and Bikrampur, most of the migrants are from Dhaka.

Settling on the char occurs through personal and social links, but the physical environment, local administration and migrants act in tacit consonance. The char dwellers quickly learn to understand and cope with the rise and fall of the river. Chars with their fertile soil and inaccessibility from mainland environment provide ideal spaces for illegal migrants to settle in. Bangladeshi farmers, with their traditional skills of farming and soil management, have converted nearly all chars of Damodar River into fertile agricultural lands.

Lalmohan describes the story of his long journey from 1952, when he arrived in West Bengal from Bangladesh, to 1962, when he bought some land in Kasba Mana to permanently settle there:

> After arriving in West Bengal from Bangladesh, I spent several months in the transitional camps in south Bengal. Then I started working as a labourer in the construction of canals in a nearby river valley project. I did not want to go and live in resettlement colonies outside West Bengal. I thus tried to change, with someone's help, the label of my past occupation in the refugee card from agriculture to business. I received 4 kathas of land to build a house and Rs 750 for setting up a business along with other families. But I was always on the look-out for agricultural land. One day I got word of some land in Char Kasba from one of my relatives. I immediately came here and bought 11 acres of land at Rs 900 in 1962. Since then I have been living here. I received the patta of these lands in 1995.

Like Lalmohan, many others also did not accept the physical relocation to unknown lands with the potential of isolation from their family relations and village friends. Many were apprehensive of the difficulties that might befall them in an environment that was totally different from the familiar alluvial floodplain of Bangladesh. Naren Sarkar of Char Bhasapur comments:

> We had no skills except farming. We didn't even know how to ride a bicycle. So how could we make our livelihoods in any other area except in the cultivable land such as these charlands? We had the skill of making land more fertile and productive. Therefore, we preferred these charlands for our resettlement and converted them into fertile crop land without any support from the West Bengal government.

Gopal Mondal of Char Gaitanpur also associates his choice of new home in charland with the availability of agricultural land. He came to India in 1972 after the Bangladesh war of independence, and settled, temporarily, in 24 Parganas from where he was looking out for cheap land. Finally he came here in 1988. He says:

> I was a farmer in Bangladesh and sold my fertile land at a negligible price before I moved to India. I saved that money for purchasing land in West Bengal. I frequently visited different places in West Bengal wherever I had the slightest contacts, searching for cheap land. I know farming better than any other work and I have the confidence of making any land productive by using methods such as manuring passed on to me by my forefathers. A relative in Udaypalli [the part of Burdwan Town across the riverbank near Char Gaitanpur] informed me of available land in this char. It was then that I purchased a large 13 acre plot for Rs 39,000.

The settlement process on the charlands intensified after the 1970s, when the Bangladesh liberation movement led to a significant increase in the flow of illegal migrants into southern West Bengal across the Bangladesh border. A large group of migrants came to settle on the charlands in 1971 following Bangladesh's independence from Pakistan. These people were mostly illegal Bangladeshi immigrants to West Bengal, who even today continue to enter India. The earlier settlers on these chars often purchased land for their relatives and neighbours, who then crossed the border illegally into India and settled here. Kinship relations, at the village and extended clan levels, had a strong influence on the shape of these new settlements. The inflow of illegal migrants and their settling of charlands is a continuing process. Our field survey revealed that, in Char Gaitanpur, 12 such families arrived between June 2000 and May 2004. Religious persecution and increasing insecurity faced by the Hindu minority in Bangladesh, together with poverty and lack of work, were cited by these people as the underlying reasons for their movement.

The illegality of the char dwellers affects their livelihood choices and coping strategies. This affects women more than men who are often in fear of visiting the markets and shops lying across the embankments. Yet, the most significant responsibilities, including providing the household food security, are commonly borne by women. These responsibilities mean they have to work should-to-shoulder with men, clearing the land, preparing it, planting seedlings or seeds for winter crops, harvesting them, as well as selling them. Women also need to buy and sell these products in the market. A difficult question is: Do the char dwellers contemplate living there permanently? Women are ambivalent about what the future holds for them and their families though being fully aware of the vulnerability of their lives. None of those we spoke to were ready to give us a concrete response – some felt that they work in the fields with their husbands in order to save money and build better lives away from this char where they would live more secure lives. At the same time, some felt that they are happier on chars than they were in Bangladesh: 'Ekhanei bhalo achhi' (we are alright here).

Charland Women

Except in partition studies, the literature on Bangladeshi migration to India had commonly been silent on the question of gender (see Datta 2004; Kabeer 1991; Kabeer and Subrahmanyam 1996; Ramachandran 2003). This is an important gap because the post-migration work of making a living generally falls on women. In the Damodar charlands, it was not different. The differences in experiences of rebuilding lives, however, are also different amongst different classes of women. Even while living on the chars many women stay rooted to their class differences as well as different social locations. These differences meant much more in their place of origin than in the chars where almost everyone works as hard as they can. The difference in income levels is more pronounced in their family's status and respect, which women tend to receive in similar ways as to the male family members.

Migration has always remained a family decision – women have no say in such matters. The participants voiced their negligible roles in the final decision to leave their ancestral homes (see Lahiri-Dutt and Samanta 2004; Samanta and Lahiri-Dutt 2007). Gandhari, for example, did not want to leave Bangladesh 'because all members of my father's family are still living there. But my husband did not listen to me.' A similar sentiment was voiced by Khukumani:

> I tried to stop my husband from coming to India then because I had two young children [one was six-months old and the other, two years]. When we migrated we often had to go without food. With two small kids, we were looking for a shelter [matha gonjar thain]. It was a very tough time for me consoling the kids when they cried out for food. We moved from one place to the other in southern Bengal for two years before coming here. Although life was not easy here, this charland has given me at least a permanent shelter.

Debirani describes her experience:

> [T]he first time I came with my husband and four children to India was in the month of June in 1984.... We spent three months during the rainy season under the cover of polythene sheets beneath a banyan tree. When it rained, we would take shelter in a verandah of another house. At that time we could not manage to get the ration cards [the most important legal document, considered the only passport of the poor migrant] and therefore went back to Bangladesh around the end of September. After two years we again left Bangladesh, and that time we came here directly as my elder son who had come here earlier had managed to establish some contacts and build a mud-hut on this char. Since then we have been living here.

Renu Das echoes a similar story. After crossing the border, she could not provide more than one meal a day for her children for at least two months – that was the worst phase in her life. But the hope of better days kept her family on the move till they found the char. In contrast to these long transitional phases that Debirani and Renu went through, some others reached the chars in relatively shorter time and in a smooth manner. Migration for them became easier because of the helping hand extended by relatives or friends who were already settled on chars. Clearly, one cannot say that the process of migration has affected the lives of women homogeneously, although one thing is clear: whatever the pathways and steps of migration, all women have had to suffer for sustaining the family after and before migration, and the char environment played a critical role in their struggle to survive (Lahiri-Dutt and Samanta 2005).

Perceptions of the Char as 'Home'

Charland women continually reconstruct their lives in their new homes along the model of what they had to leave behind. Yet, charlands begin to appear as special places with new environmental adjustments where gendered lives are interwoven with local realities and distant dreams. When asked 'How do you feel about living here?' participants reacted in different ways. Gandhari, Ashalata, and Khukumani felt that they are living well because 'I can get sufficient food and have some cash in our hand. All of my family members give me patient hearing in time of taking decision regarding any family matters.' However, some younger women revealed their discontent. Women aged above 40 with grown-up sons who did not need to work in the fields as their families were relatively better-off, were also less burdened with domestic chores as daughter-in-laws usually helped with cooking, washing, cleaning, and other housework. They did light housework and tended the domestic animals. Yet, some felt that life here was not as happy and content as it was in their homes in Bangladesh due to a lack of a wide support network to communicate and relate to.

Poorer and younger women such as Uma, Sikha, and Sandhya think they were much better-off in Bangladesh as they did not have to work outside home. Uma says:

> I have to work from sunrise to sunset without any break. I cook three times a day. Beside all my domestic chores, I work in our own field where I take my six-month-old child and put her down on the ground next to where I work.

Night is only time for rest. Sandhya says: 'Nights are too short in summer. For that reason, I prefer the winter which gives me a longer night. That way, I get a longer rest.' These women sometimes get tired of their burdensome life of charland and would prefer to settle somewhere else as soon as they can afford to (Lahiri-Dutt and Samanta 2005).

Therefore, different generations and economic groupings of char women perceive their new home differently. For some, life here is better than the violent persecution families faced in Bangladesh. To others, it provides a foothold on Indian soil that offers relatively fewer competitions from locals. The hierarchies the migrant women bring with them remain on the chars, but how long they will continue to do so cannot be determined.

Regrounding in the Chars

Glick Schiller et al. (1995: 49–50) critique the previous generation of researchers who had viewed immigrants as persons who uproot themselves, leave home and country behind, and face the painful process of incorporation into a different society and culture. In a world with rapid transport and instant communication, such complete uprootings are neither compulsory, nor necessary. As people move across national boundaries to inhabit diasporic locations, home and belonging assume different meanings, and are experienced in different ways. This ease of movement and the numbers and diversity of people who move today, have led to a significant rethinking of what comprises 'home' and how one 'belongs' to the home. Ahmed et al. (2003: 1–2), for example, note that migration does not necessarily involve a complete freedom from places, that mobility and placement are interdependent, and even grounded homes may be considered as sites of change, relocation, or uprooting. They note that the rethinking of home and migration should avoid clearly definable processes of 'migrating' and 'homing' and should consider home and migration in terms of a plurality of experiences, histories, and constituencies. As this chapter shows, this is quite the case of the Bangladeshis who made a home on chars and rebuilt their lives, and the complexities of history and contemporary politics of border-crossings in Bengal is such that any complete separation of homing and migrating yields a simplistic image of the fluid movements of people that characterise the charland communities. Banerjee (2010: 138) adds a gender angle to this emerging conceptual approach to mobility; in critiquing the universal notion of citizenship, she notes: '[T]he ideological constructions of the state are weighted against women who remain on the borders of democracy.' Indeed, as we have shown in this study, for women who are living marginal lives on the borders of legitimacy and legality, political borders do not mean much more than just a condition, a state of mind, rather than absolute boundaries that prevents mobilities effectively.

Some char women repeatedly noted that they lived 'better' in Bangladesh. Our constructions of charland lives show that not only is migration work-gendered, but that the work burdens of migrants, arriving in India without legal papers and living in an insecure environment, fall on women. Although this chapter focuses on the life experiences of a limited number of women, it enlightens us in understanding gendered processes of migration as well as in the survival techniques the women adopt to eke out a living in face of vulnerability. In migration pathways, as shown, kin relations play important roles for women. Men gave money to kin, women gave time or contacts. Apart from material exchanges, women also maintain social connections as emotional resources; migration work takes on different meanings for each class group and women work harder than men to secure their families' well-being after migration. Living on the chars equates to relentless struggle for the women of migrant families from Bangladesh, who bear the unpaid and unacknowledged work burdens in sustaining the families. It is difficult to precisely measure, in quantitative terms, the value of the work put in by the char women. This study has brought to light the variations in life experiences, workloads and seasonal variations in women's work in the char. The social construction of the char environment through the eyes of women was a particularly interesting exercise. The embeddedness of migration in social relations suggests that the social position of a woman shapes the range, meaning, and content of migration work. Women negotiate to get resources for the family shaped by their social class positions. However, women in chars perceive their work burdens and lives differently. In spite of all the risks, charlands offer a more secure environment for the migrants than that they were living in the Bangladeshi society, attributing specificity to both the environment, and the way char dwellers perceive the risks and benefits.

In tracing the regrounding of mobile but illicit subjects on non-legal lands, this chapter speaks to the growing literature on non-state-dependent forms of geographical contexts, and cross-border identities. It becomes clear that there is no clear-cut distinction between illegitimacy and the laws of the state, which are as much the product of history as mobilities that are commonly perceived as illegal. In South Asia, both the borders and the practices of illegal border-crossings have emerged from history and contemporary politics. This paper puts forth a body of evidence that contests the statist discourses of borders and boundaries, and highlights the importance of both the geographical context and human lifeworlds. The unique geographical context of chars in particular point to the need for the inclusion within mobility and migration studies of non-statist-forms of geographical units. Such units can build alternative understandings of non-state-dependent forms of location, extricate transborder mobility studies from embedded statism, and illuminate the hybrid identities of mobile subjects who cannot be pinned down geographically. Looking beyond legitimacy and illegality in flows of invisible people across visible borders, we venture into the less documented spaces that Wong (2005: 71) described as existing 'beyond the formal data-collecting gaze of the state', or the rhetorical production of boundaries through the narratives of migrants (Schendel 2005: 49). It is the micro-space of gender roles in regrounding livelihoods that literally makes the 'difference', in demolishing the conventional view of seeing migrants as persons who uproot themselves, leave home and country behind and reestablish themselves in a different society and culture (see Glick Schiller et al. 1995: 49–50 for a critique of this perspective).

Coming back to regrounding in place, we see that as people move across national boundaries to inhabit diasporic locations, home and belonging assume different meanings, and are experienced in different ways by women and men. The lives of charland people permit us to rethink what comprises 'home' and how one 'belongs' to the home, ideas that are significant for contemporary geography. Ahmed et al. (2003: 1–2) noted that migration does not necessarily involve a complete freedom from place, that mobility and placement are interdependent, and even grounded homes may be considered as sites of change, relocation or uprooting. They note that the rethinking of home and migration should avoid clearly definable processes of 'migrating' and 'homing' and should consider home and migration in terms of a plurality of experiences, histories and constituencies. As this paper shows, this is the case of Bangladeshis who made a home on chars. Banerjee (2010: 138) adds a gender angle to this emerging conceptual approach to mobility; in critiquing the universal notion of citizenship, she notes: '[T]he ideological constructions of the state are weighted against women who remain on the borders of democracy.' Indeed, as we have shown in our study, some char women repeatedly noted that they were 'better' in Bangladesh. Our constructions of charland lives show that not only is migration work-gendered, but that the work burdens of migrants, arriving in India without legal papers and living in an insecure environment, fall on women. Although this paper focuses on the life experiences of a limited number of women, it helps us understand gendered processes of migration as well as in the survival techniques the women adopt to eke out a living in the face of vulnerability. Living on the chars equates to relentless struggle for the women of migrant families from Bangladesh, who bear the unpaid and unacknowledged work burdens of sustaining their families. It is difficult to precisely measure, in quantitative terms, the value of the work put in by the char women. Our study brought to light the variations in life experiences, workloads and seasonal variations in women's work in the char. The social construction of the char environment through the eyes of women was a particularly interesting exercise. The embedding of migration in social relations suggests that the social position of a woman shapes the range, meaning and content of migration work. Women negotiate to get resources for the family shaped by their social class positions. However, women in chars perceive their work burdens and lives differently. In spite of all the risks, charlands offer a more secure environment for the migrants than those available to them in Bangladesh.

Through the narratives of their own lives and their ceaseless search for home, the mobile subjects on charlands pose important questions relating to the legitimacy of borders. In response to the state-sponsored and homogenised identities of the citizen, the char dwellers recreate and reinvent themselves as citizens of a borderless world. In regrounding themselves, the uprooted subjects contest the hegemonic view of citizenship in South Asia. The fluid nature of chars and the uncertainties of life and livelihood provide yet another border to be negotiated. In the process, conflicts with those living on the mainland as well as internal struggles and strife, ethnicity, caste and gender – all make important differences. The roles played by caste and gender appear most significant in the ways people 'settle in' or make their homes on charlands. The engendering of displacement narratives has important theoretical implications. In his call for empirical studies to embrace 'new theorisations' Boyle (2010) observes that instead of simply looking at migration as a series of push and pull factors in relation to economic forces, migrants' identities need to be recognised as factors shaping mobility and the interpretation of economic variables. Privileging women's experiences is the critical element in understanding place-making and reinvention of citizenship. Located in-between belonging and non-belonging, women are also forced to negotiate their differences with the state that denies space based on gender. This empirical study also pointed to a less-debated theoretical direction; it points to the need to explore more deeply the 'osmotic properties of place'. Geographical place is key to home-making as we locate the mobile subjects in their historical, social and political contexts; we also begin to see the important role geographical scale plays at the micro level.

References

Acharya, J.M., Manjita, Gurung, and Samaddar, Ranbir. 2003. No-where People on the Indo-Bangladesh Border, *Working Paper No. 14*, South Asia Forum for Human Rights, Kathmandu.

Adnan, Shapan. 1976. Land, Power and Violence in Barisal Village, Mimeo, The Village Study Group, *Working Paper No. 6*, pp. 1–9.

Ahmad, Imtiaz. 1997. Indo-Bangladesh relations: Trapped on the nationalist discourses, in *State, Development and Political Culture: Bangladesh and India*, edited by Barun De and Ranabir Samaddar. New Delhi. Har-Anand Publishers, 75–93.

Ahmed, Sara, Castaneda, Claudia, Fortier, Anne-Marie, and Sheller, Mimi (eds). 2003. *Introduction, in Uprootings/Regroundings: Questions of Home and Migration*. Oxford: Berg, 1–19.

Alam, Sarfaraz. 2003. Environmentally induced migration from Bangladesh to India. *Strategic Analysis:* A Monthly Journal of the IDSA, 27(3), 422–38.

Ali, S.M. 1980. Administration of char-land of Bangladesh. *Asian Affairs*, 2, 295–303.

Anderson, W. and Lee, R. (eds). 2005. *Displacements and Diasporas: Asians in the Americas*. Rutgers: University Press, New Brunswick.

Anderson, J. and O'Dowd, L. 1999. Borders, border regions and territoriality: Contradictory meanings, changing significance. *Regional Studies*, 33(7), 593–604.

Bandyopadhyay, Krishna. 2004. *'Infiltration' across the Borders: Auditing the Mainstream Media in West Bengal, WACC* [Online]'. Available at: http://www.wacc.org.uk/modules.php?name=News&file=article&sid=1596 [accessed: 10 December 2009].

Banerjee, Paula. 2010. *Borders, Histories, Existences: Gender and Beyond*. New Delhi: Sage Publications.

Barkat, Abdul and Zaman, Shafique uz. 2000. Forced outmigration of Hindu minority: Human deprivation due to Vested Property Act', in *On the Margin: Refugees, Migration and Minorities*, edited by R. Akbar Chowdhury. Dhaka: Refugee and Migratory Movements Research Unit, 25–47.

Baqee, Abdul. 1998. *Peopling in the Land of Allah Jaane: Power, Peopling and Environment, The Case of Char-lands of Bangladesh*. Dhaka: The University Press Limited.

Battistella, Graziano and Asis Maruja, M.B. (eds) 2003. Southeast Asia and the specter of unauthorized migration, in *Unauthorized Migration in Southeast Asia*. Mexico: Scalabrini Migration Centre, Quezon City, 35–127.

Bhattacharyya, Kumkum. 1998. *Applied Geomorphological Study in a Controlled River: The Case of the Damodar between Panchet Reservoir and Falta*, Unpublished PhD thesis, Department of Geography, The University of Burdwan.

Bose, Pradip Kumar (ed.). 2000. *Refugees in West Bengal: Institutional Practices and Contested Identities*. Calcutta: Calcutta Research Group.

Boyle, Paul. 2010. Ways to come, ways to leave: Gender, mobility, and il/legality among Ethiopian domestic workers in Yemen. *Gender and Society*, 24, 237–60.

Brammer, Angelika. 1994. *Displacements: Cultural Identities in Question*. Bloomington: Indiana University Press.

Buchanan, Francis Hamilton. 1828. *An Account of the District of Purnea in 1809–1810. Patna*: Bihar and Orissa Research Society.

Chakrabarti, Prafulla. 1990. *The Marginal Men. The Refugees and the Left Political Syndrome in West Bengal*. Kalyani: Lumiere Books

Chakraborty, Debesh, Gupta, Gautam and Bandyopadhyay, Sabari. 1997. Migration from Bangladesh to India, 1971–91: Its magnitude and causes, in *State, Development and Political Culture: Bangladesh and India*, edited by Barun De and Ranabir Samaddar. New Delhi: Har-Anand Publications Pvt. Ltd, 25–43.

Chambers, R. and Conway, G. 1992. Sustainable Rural Livelihoods: Practical Concepts for the 21st Century, *Discussion Paper 296*, Institute of Development Studies, Brighton.

Chatterji, Joya. 1995. *Bengal Divided: Hindu Communalism and Partition, 1932–47*. Cambridge University Press.

Chowdhury, Mahjabeen. 2001. *Women's Technological Innovations and Adaptations for Disaster Mitigation: A Case Study of Charlands in Bangladesh*,

Paper presented in United Nations Expert Group Meeting on Environmental Management and the Mitigation of Disasters, Ankara, Turkey, 6–9 November.

Costello, Celine Daly and Mahendra, Vaishali Sharma. 2001. Human rights and trafficking: Supporting women in Nepal. *Global AIDSLink*, 69(4), 1–8.

Currey, B. 1986. *Changes in Chilmari: Looking beyond rapid rural appraisal and farming systems research methods*, Paper presented in the workshop on Open Water Systems, Dhaka, 19–22 July.

Datta, Pranati. 2004. Push-pull factors of undocumented migration from Bangladesh to West Bengal: A perception study. *The Qualitative Report*, 9(2, June), 335–58.

Glick Schiller, Nina, Basch, L.G., and Blanc–Szeanton, C. 1995. From immigrant to transmigrant: Theorizing transnational migration. *Anthropological Quarterly*, 68, 48–63.

Haque, C. Emdad. 1995. The dilemma of nationhood and religion: A survey and critique of studies on population displacement resulting from the partition of the Indian subcontinent. *Journal of Refugee Studies*, 8(2), 185–209.

Harris, N. 2002. *Thinking the Unthinkable: the Immigration Myth Exposed.* London: I.B. Tauris.

Hillier, Jean and Rooksby, Emma (eds). 2005. *Habitus: A Sense of Place*. Aldershot: Ashgate.

Hugo, G.J. 2000. The crisis and international population movements in Indonesia. *Asian and Pacific Migration Journal*, 9(1), 93–129.

Human Rights Feature. 2001. Attacks on Hindu minorities in Bangladesh. Human Rights Features, *Voice of the Asia-Pacific Human Rights Network*, 48, 1–4.

Jacques, Kathryn. 2001. *Bangladesh, India and Pakistan: International Relations and Regional Tensions in South Asia*. New Delhi: Macmillan.

Jalais, Anu. 2005. Dwelling on Morichjhanpi: When tigers became 'citizens', refugees 'tiger-food'. *Economic and Political Weekly*, 23 April, pp. 1757–62.

Kabeer, Naila. 1991. *Gender, Production and Well Being: Rethinking the Household Economy*. Brighton: Institute of Development Studies.

Kabeer, Naila and Subrahmanyam, Ramaya. 1996. *Institutions Relations and Outcomes: Framework and Tools for Gender-aware Planning*. Brighton: Institute of Development Studies.

Kotoky, P., Bezbaruah, D., Baruah, J., and Sarma, J.N. 2003. Erosion activity on Majuli: The largest river island of the world. *Current Science*, 84(7), 929–32.

Lahiri-Dutt, Kuntala and Samanta, Gopa. 2004. Fleeting land, fleeting people: Bangladeshi women in a charland environment in lower Bengal, India. *Asia Pacific Journal of Migration*, 13(4), 475–96.

Lahiri-Dutt, Kuntala and Samanta, Gopa. 2005. Uncertain livelihoods: Survival strategies in charland environments in India. *International Journal of Environment and Development*, 2(2), 165–78.

Lahiri-Dutt, Kuntala and Samanta, Gopa. 2007. Like the drifting grains of sand: Vulnerability, security and adjustment by communities in the charlands of the Damodar Delta. *South Asia: Journal of the South Asian Studies Association*, 32(2), 320–57.

Levitt, Peggy and Rafael, de la Dehesa. 2003. Transnational migration and the redefinition of the state: Variations and explanations. *Ethnic and Racial Studies*, 26(4), 587–611.

Limb, Melanie and Dwyer, Claire. 2001. *Qualitative Methodologies for Geographers*. London: Arnold.

Malik, Shahdeen. 2000. Refugees and migrants of Bangladesh: Looking through a historical prism, in *On the Margin: Refugees, Migrants and Minorities*, edited by R. Akbar. Dhaka: Refugee and Migratory Movements Research Unit, 11–40.

Mallick, Ross. 1999. Refugee resettlement in forest reserves: West Bengal policy reversal and the Marichjhapi massacre. *The Journal of Asian Studies*, 58(1), 104–25.

Mohsin, A. 1997. Democracy and the marginalization of the minorities: The Bangladesh case, in *The Journal of Social Studies*, edited by B.K. Jahangir. Dhaka, Bangladesh: Centre for Social Studies, 78, 92–3.

Mukherjee, K. 2003. *India begins work on i cards for border residents*. Nation: Hindustan Times [Online], 1–2.

O'Malley, L.S.S. 1911. *Bengal District Gazetteers: Purnea*. Calcutta: Bengal Secretariat Press.

Pathania, Jyoti M. 2003. *India & Bangladesh – Migration Matrix- Reactive and Not Proactive* [Online, 17 March]. Available at: http://www.southasiaanalysis.org/%5Cpapers7%5Cpaper632.html [accessed: 29 November 2010].

Ramachandran, Sujata. 2003. 'Operation Pushback'. Sangh Parivar, state, slums and surreptitious Bangladeshis in New Delhi. *Economic and Political Weekly*, 38(7, 15 February), 637–47.

Ramachandran, Sujata. 2005. Indifference, impotence and intolerance: Transnational Bangladeshis in India. *Global Migration Perspectives*, No. 42. Geneva: Global Commission on International Migration.

Rahman, Md. Mahbubar and Schendel, Willem Van. 2003. 'I am not a refugee': Rethinking partition migration. *Modern Asian Studies*, 37(3), 551–84.

Ray, Mohit. 2009. *Illegal Migration and Undeclared Refugees – Idea of West Bengal at Stake*, Paper presented in National Seminar on Migration and Its Impact on Indian State and Democracy, Department of Politics and Public Administration, University of Pune, 13 March.

Rose, Gillian, 1993. *Feminism and Geography: The Limits of Geographical Knowledge*, University of Minnesota Press.

Roy, Tathagato. 2001. *My People Uprooted. Kolkata*: Ratna Prakashan.

Saha, Manoj Kumar. 2008. *Rahr Banglar Duranta Nadi Damodar* (in Bengali). Srirampore: Laser Art.

Samad, Saleem. 1998. *State of Minorities of Bangladesh: From Secular to Islamic Hegemony, Regional Consultation on Minority in South Asia*, Paper presented in South Asian Forum for Human Rights, Katmandu, 20–22 August.

Samad, Saleem. 2004. *State of Minorities in Bangladesh: From Secular to Islamic Hegemony, Muktamana* [Online]. Available at:

http://www.mukto-mona.com/articles/saleem/secular_to_islamic.htm [accessed: 10 December 2009].

Samaddar, Ranabir. 1999. *The Marginal Nation: Transborder Migration from Bangladesh to West Bengal.* New Delhi: Vedam Books.

Samanta, Gopa and Lahiri-Dutt, Kuntala. 2007. Marginal lives in marginal lands: Livelihood strategies of female-headed, immigrant, households in the charlands of River Damodar, West Bengal, in *Livelihoods and Citizenship in India*, edited by Sumi Krishna. New Delhi: Sage Publications, 99–120.

Scoones, I. 2009. Livelihoods perspectives and rural development, *Journal of Peasant Studies*, 36(1), 171–96.

Siddiqui, Tasneem. 2003. *Migration as a Livelihood Strategy of the Poor: The Bangladesh Case*, Paper presented at the conference on Migration, Development and Pro-poor Policy Choices in Asia, Dhaka, 22–24 June.

Skeldon, Ronald. 2002. Migration and poverty: Ambivalent relationship. *Asia Pacific Population Journal*, 17(4, December), 67–82.

Tuan, Yi Fu. 1974. *Topophilia: A Study of Environmental Perception, Attitudes and Values*. Englewood Cliffs, NJ: Prentice-Hall.

Tuan, Yi Fu. 1977. *Space and Place: The Perspective of Experience*. Minneapolis: University of Minnesota Press. (Also published by Edward Arnold, London in 1979.)

Williams, Allan and Baláž, M. Vladimir. 2002. Trans-border population mobility at a European crossroads: Slovakia in the shadow of EU accession. *Journal of Ethnic and Migration Studies*, 28(4), 647–64.

Wong, Diana (2005) The rumor of trafficking: Border control, illegal migration and the sovereignty of the nation-state, in Willem van Schendel and Itty Abraham (2005) *Illicit Flows and Criminal Things: States, Borders and the Other Side of Globalization*, Indiana University Press: Bloomington and Indianapolis, pp. 69–10

Chapter 11
Re-imagining 'Refugeehood': Reflections on Hmong Identity(ies) in the Diaspora

Roberta Julian

Globalisation, Diaspora and 'Refugeehood'

Diasporas are one possible outcome of displacement. The term 'diaspora' refers to

> … the (imagined) condition of a 'people' dispersed throughout the world, by force or by choice. Diasporas are transnational, spatially and temporally sprawling sociocultural formations of people, creating imagined communities whose blurred and fluctuating boundaries are sustained by real and/or symbolic ties to some original 'homeland'. (Ang 1994: 5 cited in Song 2003: 114)

Given current interest in the processes of globalisation and transnationalism, it is not surprising to find that the field of diaspora studies has been steadily growing over the past three decades (see, for example, Anderson 1983; Ang 2007; Appadurai 1996; Brah 1996; Werbner 1998). My own research on migrant/refugee settlement and ethnic identification can be located within this tradition. This chapter draws together some of the themes I have addressed in previous work on: *(a)* the settlement experiences of refugees and refugee women in Australia (Ganguly-Scrase and Julian 1997, 1999; Julian 1997, 1998; Julian et al. 1997), and *(b)* processes of identification among Hmong refugees and the Hmong diaspora (Julian 2003, 2004, 2004–05). Together with findings from some more recent research on the settlement of African–Australians (Campbell and Julian 2007, 2009), I draw on these themes to critically reflect on the concept of 'refugeehood' in the context of contemporary processes of globalisation.

The themes I wish to draw on are:

1. The problems that arise from the false dichotomy that identifies refugee movements as involuntary and other types of migration as voluntary. This dichotomy tends also to coincide with another false dichotomy, namely, that most migration is 'caused' by economic factors, but that refugee movement is 'caused' by political factors. Richmond (1988) argued the falsity of these distinctions as early as in the 1980s. In arguing for a comprehensive theory of migration that can account for both 'types' of movement, he emphasised that in the context of contemporary global systems, the interdependence

of economic, social, and political factors in all population movements must be recognised. Even then he pointed out that 'it is no longer possible to treat "refugee" movements as completely independent of the state of the global economy (1988: 12). This theme has been emphasised in more recent analyses of international migration (see, for example, Castles 2003).

2. The importance of critiquing the public image of refugee women as invisible and powerless victims. In Australia, refugee women rarely represent themselves in any public forum. Typically, they are spoken about and represented by others, such as feminist academics and migrant women of longer standing in Australia, many of whom are concerned to publicise the plight of refugee women. However, this political project depends in part upon a process of 'victimisation'. Thus, in the public sphere, refugee women are predominantly represented as victims: of war, of persecution, of tragedy, and of state and patriarchal oppression. It is important, however, to recognise the agency of refugee women and to analyse where, when, and how it occurs (see Julian 1997).
3. The significance of 'refugee' status in terms of lived experiences. Refugee women experience many of the same tensions and contradictions as other migrant women. However, they face additional pressures as refugees and as recent arrivals, and are even more likely to be represented through the discourse of victimisation. The 'refugee problem' is often represented as a Third World problem, a problem arising out of conflicts in 'other' societies. Acceptance as a refugee is predicated on successfully presenting oneself as a 'victim' of global economic and political processes. Within this context, refugee women are typically represented as the most powerless of the 'outsiders' accepted into the country. Like migrant women, they are viewed as dependents of their husbands, and like 'Third World' women, they are represented as non-modern, that is, as oppressed, disempowered victims (Morokvasic 1988). This representation sits comfortably with the emancipatory discourse adopted by many who study and work with migrant and refugee women. This modernist hegemonic discourse, however, homogenises and stereotypes all refugee women as victims. Defining these subjects as 'refugee women' denies, first, the relevance of their experiences prior to be accepting as 'refugees' and, second, the diversity of experiences which characterises the members of the category. While 'refugee women' may share an on-arrival status as marginalised and powerless, the diversity of their life experiences prior to this point in their life histories is significant. Diverse social locations in their countries of origin provide the bases for diverse biographical trajectories as women 'become' refugees. These may vary significantly among refugees from the same country of origin depending, for example, on class background, religion, age, marital status, and place (for example, rural or urban location).It is precisely because of the complex interplay of economic and socio-political factors creating the context for the existence of refugees that 'refugee women' are even less

likely to represent an homogenous category than 'migrant women'. The experiences of refugee women in their homelands are clearly an important factor influencing their settlement strategies. Since the way in which they became refugees varies dramatically, the meanings they attach to their status as refugees will vary, as will the ways in which they respond to opportunity structures in the country of resettlement.

4. The need to problematise the assumption of 'fixity' and identification with place (especially the nation state) in much of the migration and refugee literature. While this assumption is predominant in much of the migration literature, it can be argued that most populations throughout history are marked by 'mobility' rather than 'fixity' – the idea that people essentially stay put. For example, Kleinschmidt (2003) argues that we should broaden our thinking by recognising that migration has been a recurring feature in human history (though not necessarily measurable in terms of state-recognised 'immigrants' and 'emigrants'), and that we often base our work on a 'residentialist fallacy' asking questions about why people move rather than thinking about why people stay put.

The following discussion of Hmong refugees and identification processes in the Hmong diaspora provides the empirical basis for these reflections. The analysis is based on almost 10 years of association with Hmong people – mainly in Australia and the United States (US), but also in Thailand. This involved a wide variety of methods: observation, participant observation, in-depth semi-structured interviews, a survey and structured interviews, document analysis (including books, newsletters, articles on the Internet, and keynote addresses at conferences), analysis of scholarly publications (by Hmong and non-Hmong), autobiographies and biographies of Hmong (such as students and women), cultural artefacts and other media (for example, storycloths or *paj ndau*), plays (for example, *Highest Mountain Fastest River*), museum displays (for example, 'Hmong of the Mountains'), comedy skits and videos (for example, Tou Ger Xiong's 'Hmong Means Free'), and poetry and short stories by Hmong-Americans (for example, in the journal *Paj Ntaub Voice* and the newsletter *Hnub Tshiab*). In short, the research has been ethnographic and longitudinal. This chapter draws particularly on my experiences of participant observation. My aim is to subject the discourses of Hmong identity to sociological analysis, thereby revealing some important characteristics of this displaced population.

The Hmong Diaspora: A Transnational Community?

Who are the Hmong? This is a surprisingly common question about the members of a relatively small but widely dispersed diaspora. As with all questions that address issues of identity, the answer is complex and highly contested. As Stuart Hall has argued, identities are *positionings* (1990: 53). They are dialogic, highly contextual, and inherently political. Identities are not free-floating. They are territorially grounded lived experiences and as such they are constructions in progress. By examining the nexus between Hmong resettlement and identification processes in a global context, this chapter demonstrates that Hmong identities are grounded in both global and local social relations. The analysis suggests that the diaspora and the nation state are equally significant in the construction of contemporary Hmong identities.

The Hmong are one of the minority groups in Laos known as 'hill tribes'. As one of the two main ethnic groups living in the highest mountain regions in Laos, they have a minority status vis-à-vis the dominant group, the lowland Lao (Chan 1994). During the Indo-Chinese War, the Hmong supported the Americans, many working directly for the Central Intelligence Agency (CIA). In response to persecution by the Lao government, many were forced to move into the jungle, becoming refugees in their 'own' country before escaping to refugee camps in Thailand (Dean 1993). Since 1975, the Hmong have been accepted as refugees in countries such as the US, Australia, France, Canada, and French Guiana.

The existence of transnational connections among the Hmong who have resettled in these new countries suggests they constitute what Faist (2000) has termed a 'transnational social space'. For him (2000: 309):

> … transnational social spaces are relatively permanent flows of people, goods, ideas, symbols, and services across international borders that tie stayers and movers and corresponding networks and non-state organizations; regulated by emigration and immigration state policies.

The concept of transnational social space emphasises the existence of social networks and organisations 'in at least two geographically and internationally distinct places' (Faist 2000: 197). This opens up the space for a comparative analysis of the distinctive characteristics of settlement in different nation states. Thus, rather than suggesting a denationalisation process as some globalisation theorists have done (see, for example, Cohen 1997), Faist sees transnationalisation as occurring alongside the maintenance of the nation state (2000: 200):

> The transnational social spaces inhabited by immigrants and refugees and immobile residents in both countries thus supplement the international space of sovereign nation-states.

One type of transnational linkage that can emerge in such a space is a transnational community; that is, a social grouping characterised by a high degree of social cohesion, moral commitment, and continuity over time, but one that cuts across one or more nation states (Cohen and Kennedy 2000: 375). My analysis of Hmong in Australia and the US indicates that the Hmong constitute a transnational community. Social ties, especially clan ties, are mobilised between Australia and the US and identification as Hmong provides a basis of solidarity across this transnational social space (Faist 2000). For Lemoine, the French anthropologist who has conducted research on the Hmong for over 40 years, such a finding would not be surprising. His research demonstrates that mobility has characterised the Hmong way of life for an extensive period of time and that kinship relations constitute the foundation of Hmong culture and identification. As he notes (2008: 10):

> Overwhelmingly, 40 years ago, Indochinese [H]mong were not peasants but tribal farmers in the hills. They did not care for landownership and freely wandered in the forested mountains where they carved their swidden. When the land they exploited became too poor, they freely moved to another location they had previously selected.… A [H]mong village was never a permanent community and could split to join other villages or form a new settlement with other villagers. Only kinship relations, that is: lineage and clan ties were consistent and were the guidelines for close or far away migrations.

Transnational linkages exist between Hmong in Asia and in countries of resettlement (Schein 1998, 2000; Tapp 2004). Importantly, Tapp and Lee (2004) suggest that these linkages are stronger between Hmong in countries of resettlement in the First World (for example, between Australia, the US, and Canada) than they are between Hmong in First and Third World countries. Research in French Guiana, for example, found that:

> … [t]ies between Hmong in the U.S. and French Guiana are strong, with relatives visiting each other or communicating by phone. Guianese Hmong are aware of New Year celebrations, beauty pageants, and politics in the U.S. (Clarkin 2005: 5–6)

Two important aspects of the Hmong transnational community should be noted. First, perceptions of the transnational dimension vary according to social location. As a generalisation, the Hmong in Australia are more likely to be aware of, and have regular contact with, those in the US. The Hmong in America are more likely to have a vague awareness of the existence of those in Australia; although many of the second generation do not even have this. It has also been noted that 'many Hmong in the US are unaware of the fact that there are Hmong in tropical South America [that is, French Guiana]' (Clarkin 2005: 1). Thus, the topography of the transnational community varies in accordance with geographical location.

Second, the members of this transnational community, although united by a sense of shared identity, express their 'Hmongness' in distinctive ways. The Hmong women in Tasmania and the Hmong women in St. Paul clearly live their Hmong femininity differently. As Brah (1996) notes, diasporas are no more unified and homogenous than cultures or communities are. They are 'lived and re-lived through multiple modalities' as 'differentiated, heterogenous and contested spaces, even as they are implicated in the construction of a common "we" ' (Brah 1996: 184).

The Hmong Diaspora

Significant class and status differences are apparent in the Hmong diaspora. The first Hmong refugees consisted of the political and military elite who were able to exploit pre-war connections with politicians and educators in France or else manage a CIA-sponsored airlift from the military base at Long Cheng. Among those who escaped to a refugee camp in Thailand, those initially selected for resettlement went to the US. These refugees were the most educated and those who had held higher positions within General Vang Pao's army. By 2002, there were 186,000 Hmong refugees and their descendants residing in the US (Hmong Resource Center 2002). The major centres of Hmong settlement are in the mid-West (St. Paul in Minnesota and in Wisconsin) and in California (especially Fresno in the Joaquin Valley and Sacramento, the state capital). The largest concentration of Hmong in the world is located in St. Paul. Considerable variation exists among Hmong-Americans with respect to socio-economic status. There is an identifiable elite which has achieved educational and material success as well as a larger percentage which is welfare-dependant and struggling economically (Pfeifer and Lee 2004). A very high proportion of Hmong-Americans are Christians, many having converted to Christianity during their period of time in the refugee camps in Thailand.

By contrast, the Hmong community in Australia is more homogenous. The majority were 'foot soldiers' in General Vang Pao's army, and most arrived after spending a number of years (up to 15 years) in Thai refugee camps. The majority spent most of their time at the same refugee camp, Ban Vinai, which was closed in the mid-1990s. Australia is now home to approximately 2,000 Hmong (Lee 2004: 14). The centres of Hmong settlement in Australia have been Sydney, Melbourne, and Hobart with a demographic shift in recent years to Queensland sites, namely, Innisfail, Cairns, and Brisbane. The overwhelming majority of Hmong in Australia have retained their traditional religious beliefs and practices; only a very small minority have converted to Christianity. There is evidence that the Hmong in Australia are viewed as more conservative than those in other countries of resettlement such as the US and France (Tapp and Lee 2004). Recent research on the Hmong in Australia has explored the exodus of the majority of the Hmong who settled in Hobart during the 1980s and 1990s to various locations in Queensland during 2002–2003 (Eldridge 2008). Eldridge (2008: ii) discusses this movement in terms of secondary migration and concludes that:

> … in addition to the desire to create a mega-community of Hmong in Queensland – in an attempt to counter loss of tradition and culture, and build Hmong cohesiveness – secondary migration was influenced by a desire for family reunification and a strong economic motive.

Similar movements have been identified in the US where, it has been argued, they are fundamentally the same as traditional Hmong relocations that are motivated by economic considerations (Thao 1982).

Identity Narratives in the Hmong Diaspora: Unity and Diversity

The Global Discourse

When I originally began exploring the question 'Who are the Hmong?' I encountered a highly visible, well-articulated, and almost *unitary* narrative. My interviews with the Hmong in Australia and the US, together with my reading of books and videos, identified the same story about Hmong identity. This identity narrative is a hegemonic discourse emanating from Hmong-America which is subsequently diffused throughout the more peripheral points of the diaspora. The diasporic public sphere is, thus, dominated by the voices of (predominantly male) Hmong leaders in the US. Gary Yia Lee, an Australian Hmong with a PhD in anthropology, articulates the call for a unified Hmong identity that dominates this narrative (1996: 13-14):

> The Hmong, no matter where they are, need to know that the total sum is always bigger than its parts: the overall global Hmong identity is greater than its many local differences and groups... The biggest challenge for all Hmong is ... to turn our diverse language and customs into one unified and one Hmong/Miao identity...

This global-identity construction is made possible by the existence of clan ties across nation states together with technological advances in communication (such as the Internet) that enable the maintenance of social ties and the transmission of information on a global scale (Gorman and McLean 2003). However, rather than seeing this unitary narrative as representing internal homogeneity based on 'authenticity', I wish to follow Werbner's (1998) lead in asserting that this is the outcome of limitations in the diasporic public sphere. As the public sphere expands in time and space, other versions of the Hmong identity can be discerned.

The dominant identity narrative in the global discourse on the Hmong identity can be described as the quintessential 'refugee story'. It is a 'heroic' narrative that incorporates a number of themes: the war and the military; exodus and the refugee experience; and continuity with the past through a valuing of clan ties, 'traditional' Hmong culture, and memories of a 'lost' homeland. This identity narrative is located predominantly in the past. While it acknowledges the future (through, for example, the value of educational achievement in the country of resettlement), it does so in the context of a trajectory emanating from the past. The dominant leitmotif is that of 'the refugee'.

In this dominant narrative, Hmong identity is forged on a past relationship with the US in Laos. It is often articulated in books about the Hmong in America, as illustrated in the following lines, based on the translation of an interview with Chia Koua Xiong, a Hmong man: Who are the Hmong? ….

> In Laos, we helped you fight the war. The Americans came to live with our leaders in our country.… We provided food.… If the Americans came to our house, whatever we ate we treated the Americans equally.… If we found an injured soldier … we … carried the American to the base.… In some dangerous situations we were willing to let ten Hmong soldiers die so that one of your leaders could live …. We considered Americans as our own brothers.… Now we have lost our own country…. (Pfaff 1995: 7).

Numerous videos and books recount the same story of the Hmong exodus from Laos to resettlement in the West: first, of a 'brotherhood' with US citizens as a consequence of fighting side-by-side in the Vietnam War. This unique affiliation is extended to America's allies in the 'West', especially to Australians. The Hmong in Australia see themselves as veterans of the Vietnam War and it is from this base that they project their future. In the words of a Hmong man in Hobart:

> [l]ots of plans for the future. I want the Australian government to recognise that we were part of the contingency for the Vietnam War and I want our people to march on Anzac Day as part of that group. I want us to be able to practise our rituals openly and share it with everyone rather than behind closed doors.… I also want all my people to be self sufficient and independent of social welfare.

The traumatic experiences associated with exodus and flight are equally important components of this dominant narrative. For the majority of Hmong refugees, the flight from Laos was a communal affair. Whole villages and extended families lived in exile in their own country before making their way across the Mekong River into Thailand. A Hmong man in Hobart recounts:

> At the time we travelled [there were] quite a lot of people. About more than 100 people travel … from Laos to Thailand.… Many villages want to travel at the same time. So quite a lot of people.… I remember some … families, they've got very young children. About two, three years old. And when they reach the very danger[ous] place, for example very close to the communist position, they cannot let any children cry or make noise. So when they cry they do like that, you know [the interviewee gestures by putting his hand over his mouth]. And sometime children die, you know … because they cannot breathe. Some children, they put … some medicine [that is, opium] to make them sleep, and sometime they say 'Oh, that child is very tired now'. But they got too much medicine and sometime children die. It is happen … very bad.

Public events involving the Hmong almost always contain accounts of refugees enduring forced migration. Typical is the following keynote speech, from the Fourth Annual National Hmong Conference held in Denver, Colorado, in April 1998. Dr Mymee Her, a young Hmong woman, a psychologist in California, declares:

> The Hmong are classified as REFUGEES.... Hmong refugees come to the United States wounded. Most have been beaten up physically and emotionally. They seek out shelter from whatever country will offer them safety. They have no anticipation of what life holds for them in the country of refuge. They are in a state of shock, not realizing what had just happened. (Her 1998)

While this dominant narrative is predominantly articulated by male elites, it is also reproduced by women. This is most apparent in the storycloth (*paj ndau*), a form of needlework first made around 1976 in Ban Vinai refugee camp (Anderson 1996: 30). Commonly, the storycloths chronicle village life in the mountains of Laos, depict religious ceremonies, or illustrate Hmong folktales. The majority created in the 'West' or for the Western market, however, recount the transition from village life prior to the war, through the escape from Laos after the war, to life in refugee camps in Thailand, and often 'end' with an image of the plane in Bangkok that was to take them to 'freedom' in the US. As Anderson (1996: 28) states:

> The storycloths are a link with the past. They are shared memories captured in visual images, with the embroiderer's needle rather than the camera or the written word. As episodes of social history, they record and pass on information about Hmong customs to the younger generation, especially those born in the United States with no first-hand knowledge of Laos.... They are a form of non-verbal communication that transcend [*sic*] language barriers.

While it is highly visible in the US, this diasporic identity is also evident among the Hmong on the diaspora's periphery, particularly in Canada, France, and Australia. The boundaries of this 'imagined community'[1] (Anderson 1983), though, are not spatial; they are grounded in social networks that constitute the diaspora. This is evidenced by the fact that this global narrative resonates with some Hmong in Bangkok (who communicate regularly on the Internet with the Hmong in the US) and in Wat Tham

1 The term 'imagined community' refers to the fact that, typically, nations (and other types of communities) are conceived as homogenous despite the existence of diversity and inequality within their boundaries (Holmes et al. 2003: 498).

Krabok[2] (where Hmong refugees are in receipt of financial remittances from the Hmong in the US), but not with Hmong hill-tribes in northern Thailand.[3]

The narratives recounted here support and reinforce an essentialist notion of Hmongness and encourage the maintenance of ideas of 'tradition' and 'authenticity'. Those in the West attempting to reconstruct 'tradition' look to Thailand, Laos, and even China for 'authenticity' (Schein 1998). For many Hmong, visits to these countries are viewed with enthusiasm and valued as a way of reclaiming traditional Hmongness. They serve to reconnect the members of the transnational community with a 'homeland' and, thus, further strengthen the unity of the diaspora. It can be argued that the maintenance of such connections with the past involves a *reinvention* of tradition which takes the form of a 'strategic essentialism' (Ang 1993; Spivak 1990). This involves the mobilisation of essentialist views of identity – as inevitable, fixed, and stable over time – for strategic, although not necessarily conscious, purposes.

Voices of Resistance

It is possible to identify resistances and challenges to the global hegemonic discourse identified in the previous section. Such resistances arise where the narrative of the 'refugee success story' does not resonate with lived experience. These resistant voices arise in local contexts and reflect differences within the Hmong diaspora based on gender, age, class, religion, and place. They are what Werbner (1998: 12) refers to as voices of 'argument and imaginative creativity.'

The voices of resistance are predominantly those of educated Hmong women in the US and France. However, they have been joined more recently by the voices of Hmong youth, both male and female, throughout the diaspora. Thus, these voices demonstrate the significance of age, generation, and gender as bases of diversity in the transnational community. The main topics of contestation and debate include clan structure, youthful marriage, cross-cultural marriage, polygamy and levirate, 'kidnap' marriages, women's education, and the significance of Hmong language for Hmong identity.

Hmong women play a leading role in the process of 'translating' Hmongness as they resist, challenge, and negotiate new constructions of Hmong femininity (Donnelly 1994). An emerging reflexivity is increasingly apparent in these strategies as they become articulated in various media. For example, in 1995, a small group of young Hmong women in the US organised themselves as the 'Hmong Women Educational Delegation' or HWED to the 4th United Nations

2 Wat Tham Krabok (the last of the Hmong refugee camps) was established within the boundaries of a Buddhist temple in Thailand. It was closed in 2004 and many of those who had lived there were relocated to the US (Grigoleit 2006).

3 Personal communication with Ralana Maneeprasert of the Tribal Research Institute, Chiang Mai, Thailand, 2001.

World Conference on Women held in Beijing (as representatives of Hmong women throughout the world). Their brochure stated the following:

> After close to twenty years in the U.S., the majority of Hmong women have, for the most part, survived the initial culture shock, and are adapting well to the diverse American cultures. The many opportunities available in the U.S. have been especially beneficial to Hmong women. Each year a growing number of women receive formal education and are finding employment. Beyond family responsibilities they are actively involved in their communities at all levels and in different capacities. They have taken the initiative to promote the advancement of women through economic self-sufficiency and to advocate for Hmong women's representation in leadership positions in the educational, political, and social sectors. In their new roles, many Hmong women find that in order to positively impact their communities, they must obtain leadership skills that are not only sensitive to their culture but also relevant to their gender. (HWED 1995)

New communications media, such as the Internet, provide a means of reaching a global Hmong audience. There is an online Hmong journal and at least one Hmong homepage in both the US and Australia. Young Hmong people in the US and Australia discuss issues surrounding Hmong identity in various chat rooms.

Hmong women are creating spaces for the articulation of resistances to the hegemonic narrative. In 1994, Mai Neng Moua launched a Hmong literary journal, *Paj Ntaub Voice* which is subtitled *A Journal Giving Expression to Hmoob Voices*. Since then she has published an anthology of Hmong American literature (Moua 2003a). The publishers' description of the book is telling:

> In this groundbreaking anthology, first- and second-generation Hmong Americans – the first to write creatively in English – share their perspectives on being Hmong in America.... These writers don't pretend to provide a single story of the Hmong; instead, a multitude of voices emerge, some wrapped up in the past, others looking toward the future, where the notion of 'Hmong American' continues to evolve. (Moua 2003b)

The oppositional voices of Hmong women and young people are also being heard in other arenas such as theatres and art galleries. In Minneapolis, Minnesota, 'Theater Mu' has become an avenue for Hmong actors and Hmong plays, and in April 2002, an alternative art gallery held an exhibition of works by Hmong-American artists. In addition, the field of popular culture is a newly emerging and extremely significant site for the expression of these voices. At the 2002 Hmong National Development Conference Youth Forum in Milwaukee, a wide range of popular cultural acts were on display – from break dancing to poetry – with young Hmong women well-represented.

Importantly, these resistant and oppositional positionings (see Hall 2002 [1976]) are collective strategies. Gigi Durham (1999) has criticised the dominant views of resistance among young women in that they typically adopt a model of autonomous individuals. She argues that this overlooks 'the crucial role of women's relationships with other women in their constructions of social reality' (Gigi Durham 1999: 215). Many of the discussed strategies demonstrate the successful translation of the Hmong identity through the collective activities of the Hmong women. Significantly, these resistant readings are also informed by 'other' discourses including feminist analyses of women of colour and Asian-American narratives and practices. Thus, as Hein (1994) notes in the context of US race relations, the Hmong identity has shifted from that of 'migrant' to 'racial minority'. For example, Tou Ger Xiong, a freelance dramatic artist and stand-up comedian, uses rap to narrate and 'translate' Hmongness in a way that creates an identification with African-Americans. His message of Hmongness is as follows (Xiong 1998):

> As you can see I'm Asian; yeah, I'm not black.
> What I'm about to say might sound like slack,
> But just lend me your ears, and hear me out,
> I've come to tell you what I'm all about.
> Yes my name is Tou, and I've come to say
> that I'm special, talented in many ways.
> Yes I know kung fu, and martial arts,
> You try to go get me, man I'll tear you apart.
> Yeah I'm bad, mean, and tough is my game.
> They call me the master; yes, it's my name.
> Well you might think it's weird, to see that I'm Asian,
> Bussing some rhymes on such an occasion.
> Well let me tell you, how I came to be,
> I was born in Laos in seventy-three.
> In my culture, we sing and dance,
> But I'm a style rapper, yeah I take my chance.
> Yo, even though this, is my first rap,
> You don't have to like it, and you don't have to clap.
> To those who listen, it might be nice,
> To see this Hmong boy, kickin' like Vanilla Ice.

Alternative Hmong Identities – Translating Global Discourses in Local Contexts

Comparative analysis is a powerful tool. The differential trajectories of Hmong identities in Australia and the US demonstrate the existence of both a *global* narrative of Hmong identity and the construction of various Hmong identities in a *transnational* social space. The global narrative offers a sense of unity in the diaspora while simultaneously providing the base material for resistance and/or 'translation' in different local contexts. In the US, Hmong women and youth have extended the diasporic public sphere by creating spaces within which to voice resistant readings of Hmong identity. In doing so, they act as powerful of agents of change. The strategies they have chosen look towards the future; they anticipate the survival of Hmong identity in the West via its translation.

It is important to recognise that these young women and youth are predominantly the children and grandchildren of the educated political elite who were among the first Hmong to flee Laos and resettle in the US. Thus, class must be identified as a significant factor in the selection of strategies and identities insofar as it structures access to resources, networks, and opportunities in countries of resettlement. The Hmong in the US are differentiated along class lines with a significant proportion that is welfare-dependant and living below the poverty line (Pfeifer and Lee 2004). These Hmong have been identified as less well-integrated in the new country than those who are more educated and economically successful (Hein 2006). Many were members of General Vang Pao's army in Laos, were still fiercely loyal to him up until his recent death and they look towards repatriation in Laos as the way to a more secure economic situation. Their narratives have not been so clearly articulated in the public sphere.

The Australian context is quite different. The repertoire of adaptive strategies (including gender strategies) available to the Hmong is more restricted. As a very small and dispersed population with generally low socio-economic status, the Hmong in Australia have limited resources to draw upon. For Hmong women in Australia, the global narrative provides the dominant 'cultural notion of gender at play' (Chen 1999: 589). At the level of the nation state, the rhetoric of multicultural policy in Australia 'encourages' the presentation of Hmong identity. Typically, however, multicultural spaces in Australia are aimed at the white middle-class or the tourist market – the identities presented are essentialist and 'traditional'. As Hage (1998) argues, they are limited to representations that sustain 'fantasies of white supremacy' among the dominant majority. In Australia, multicultural policy thus reinforces the power of the global narrative; it does not offer spaces for 'translation'. Consequently, while in the US, resistance takes the form of 'translation' and reinvention in 'the third space' (Bhabha 1990), the lack of such spaces in Australia leads to 'a wholesale rejection of, or disinterest in, Hmong cultural traditions' (Tapp 2004).

Given Australia's ambivalent relationship with Asia and 'Asians' (Ang 2000), spaces for the creation of hybridity are limited. They include an emerging media in the field of popular culture (see Ang et al. 2000), but the Hmong have limited access to these. Nevertheless, there are two such spaces available to the Hmong. Significantly, both are grounded in the transnational social space. One is the Internet and the other is the transnational 'community' of Hmong women. First, the Internet offers a space within which the Hmong in Australia can engage in the translation of Hmong identity. It is used by Hmong youth and is dominated by the Hmong in the West (Tapp 1994). Second, Hmong women who have made their mark as 'trailblazers' (Bays 1994) in the US have become increasingly aware of the situations of Hmong women in Australia. Through e-mail communication and visits to Australia, they have begun to establish transnational ties. Importantly, these ties are not clan-based: they are *gendered* ties between Hmong women. Educated Hmong women in the US are offering themselves and their experiences as resources for the translation of Hmong femininity in Australia. The ultimate goal, however, is to contribute to the survival of the Hmong identity in the diaspora.

Conclusion

There has been plethora of studies on globalisation in the last two decades (for example, Hall 1992; Rex 1995; Robertson 1992; Schirato and Webb 2003). Globalisation is commonly understood to refer to a complex set of social, economic, political, and cultural processes that cut across national boundaries and increase levels of interconnectedness (Holmes et al. 2003: 497). Many of these studies, however, do not adequately distinguish between the processes of globalisation and transnationalisation. Some globalisation scholars (for example, Appadurai 1996; Harvey 1989) have emphasised the 'detached nature of cultural representations in global flows' (Faist 2000: 210). In contrast, transnationalisation is not deterritorialised; it refers to processes that are grounded in space and place, particularly the nation state. As Faist (2000: 211) notes:

> Whereas global processes are largely decentred from specific nation-state territories and take place in a world context, transnational processes are anchored in and span two or more nation-states, involving actors from the spheres of both state and civil society.

This chapter has demonstrated that global narratives of Hmong identity must be seen as analytically distinct from, but empirically intertwined with the constructions of Hmong identities in transnational social spaces. The latter are grounded in the opportunities and constraints of the local contexts that are framed by economic, political and social processes in the nation state.

The Hmong diaspora is differentiated along the lines of class (and/or status), religion, age, generation, gender, and, most importantly, place. Adaptive strategies are selected and identities are constructed in the context of the policies and institutional practices of the nation state as well as the diaspora. These different terrains offer quite different sets of opportunities, constraints, and resources from which individuals can 'choose' their own identities (Song 2003). This leads to contestation and debate over the meaning of Hmongness in the diaspora.

Theories of displacement must be informed by studies of displaced populations that are grounded in specific contexts. This chapter has highlighted the value of conducting longitudinal and comparative ethnographic research to ensure that the complex contours of the effects of displacement are illuminated and the contested nature of identification is revealed as people select settlement strategies and develop a sense of 'belonging' in their new homes.

References

Anderson, Benedict. 1983. *Imagined Communities*. London: Verso.

Anderson, June. 1996. *Mayko's Story: A Hmong Textile Artist in California*. San Francisco: California Academy of Sciences.

Ang, Ien. 1993. The differential politics of Chineseness, in *Community/Plural 1/1993 – Identity/Community/Change*, Ghassan Hage and Lesley Johnson. Sydney: Research Centre in Intercommunal Studies, University of Western Sydney, 17-26.

Ang, Ien. 1994. On not speaking Chinese. *New Formations*, 24 (Nov.), 1-18.

Ang, Ien. 2000. Asians in Australia: A contradiction in terms? in *Race, Colour and Identity in Australia and New Zealand*, edited by John Docker and Gerhard Fischer. Sydney: UNSW Press, 115–30.

Ang, Ien. 2007. Beyond Asian diasporas, in *Asian Diasporas: New Formations, New Conceptions*, edited by R.S. Parrenas and C.D.S. Lok. Palo Alto, CA: Stanford University Press, 285–90.

Ang, Ien, Sharon Chalmers, Lisa Law and Mandy Thomas (eds). 2000. *Alter/Asians: Asian-Australian identities in Art, Media and Popular Culture*. Sydney: Pluto Press.

Appadurai, Arjun. 1996. *Modernity at Large: Cultural Dimensions of Globalization*. Minneapolis: University of Minnesota Press.

Bays, Sharon Arlene. 1994. Cultural Politics and Identity Formation in a San Joaquin Valley Hmong Community. Unpublished Ph.D. dissertation, Department of Anthropology, University of California, Los Angeles. UMI Order Number 9513829.

Bhahba, Homi. 1990. The Third Space: Interview with Homi Bhahba, in *Identity: Community, Culture, Difference*, edited by J. Rutherford. London: Lawrence and Wishart, 201-21.

Brah, Avtah. 1996. *Cartographies of Diaspora: Contesting Identities*. London: Routledge.

Campbell, Danielle and Roberta Julian. 2007. Community policing and refugee settlement in regional Australia: A refugee voice. *The International Journal of Diversity in Organisations, Communities and Nations*, 7(5), 7–16.

Campbell, Danielle and Roberta Julian.2009. *A Conversation on Trust: Community Policing and Refugee Settlement in Regional Australia*, ARC Linkage Grant Final Report to Industry Partners, Tasmanian Institute of Law Enforcement Studies, University of Tasmania.

Castles, Stephen. 2003. The international politics of forced migration. *Development*, 46(3), 11–20.

Chan, Sucheng. 1994. *Hmong Means Free: Life in Laos and America*. Philadelphia: Temple University Press.

Chen, Anthony S. 1999. Lives at the center of the periphery, lives at the periphery of the center: Chinese American masculinities and bargaining with hegemony. *Gender and Society*, 13(5), 584–607.

Clarkin, Patrick F. 2005. 'Hmong resettlement in French Guiana. *Hmong Studies Journal*, 6, 1–27.

Cohen, Robin. 1997. *Global Diasporas: An Introduction*. London: University College London Press.

Cohen, Robin and Paul Kennedy. 2000. *Global Sociology*. London: Macmillan.

Dean, Elizabeth. 1993. The Hmong in Hobart. *Post Migration*, 91(April), Canberra: Department of Immigration and Ethnic Affairs, 17–18.

Donnelly, Nancy D. 1994. *The Changing Lives of Refugee Hmong Women*. Seattle: University of Washington Press.

Eldridge, Margaret. 2008. *New Mountain, New River, New Home? The Tasmanian Hmong*, Unpublished Master of Arts dissertation, School of History and Classics, University of Tasmania.

Faist, Thomas. 2000. *The Volume and Dynamics of International Migration and Transnational Social Spaces*. Oxford: Oxford University Press.

Ganguly-Scrase, Ruchira and Roberta Julian. 1997. The gendering of identity: Minority women in comparative perspective. *Asian and Pacific Migration Journal*, 6(3–4), 415–38.

Ganguly-Scrase, Ruchira and Roberta Julian. 1999. Minority women and the experiences of migration. *Women's Studies International Forum*, 21(6), 633–48.

Gigi Durham, Meenakshi. 1999. Articulating adolescent girls resistance to patriarchal discourse in popular media. *Women's Studies in Communication*, 22(2), 210–29.

Goffman, Erving. 1961. *Asylums: Essays on the Social Situation of Mental Patients and Other Inmates*. New York: Anchor Books.

Gorman, Lyn and David McLean. 2003. *Media and Society in the Twentieth Century*. Malden, MA: Blackwell.

Grigoleit, Grot. 2006. Coming home? The integration of Hmong refugees from Wat Tham Krabok, Thailand, into American society. *Hmong Studies Journal*, 7, 1–22.

Hage, Ghassan. 1998. *White Nation: Fantasies of White Supremacy in a Multicultural Society*. Sydney: Pluto Press.

Hall, Stuart. 1990. Cultural identity and diaspora, in *Identity: Community, Culture, Difference*, edited by J. Rutherford. London: Lawrence and Wishart, 22–37.

Hall, Stuart. 1992. The question of cultural identity, in *Modernity and Its Futures*, edited by S. Hall, D. Held, and T. McGrew. Cambridge: Polity Press in association with the Open University, 273–326.

Hall, Stuart. 2002 [1976]. The television discourse: Encoding and decoding, in *McQuail's Reader in Mass Communication Theory*, edited by Denis McQuail. London: Sage Publications, 302–08.

Harvey, David. 1989. *The Condition of Postmodernity*. Oxford: Blackwell.

Hein, Jeremy. 1994. From migrant to minority: Hmong refugees and the social construction of identity in the United States. *Sociological Inquiry*, 64(3), 281–306.

Hein, Jeremy. 2006. *Ethnic Origins: The Adaptation of Cambodian and Hmong Refugees in Four American Cities*. New York: Russell Sage Foundation.

Her, Mymee. 1998. *Hmong Resiliency: Surviving a War and Living a Dream*, Keynote address given at the Fourth Annual Hmong National Conference 'Living the Dream', Denver, Colorado, USA, 16–18 April.

Hiruy, Kiros. 2009. *Finding Home Far Away from Home: Place Attachment, Place-identity, Belonging and Resettlement among African-Australians in Hobart*, Unpublished Master of Environmental Management thesis, School of Geography and Environmental Studies, University of Tasmania.

Hmong Resource Center. 2002. *Newsletter*, No. 10. St. Paul, MI: Hmong Resource Center.

Hmong Women Educational Delegation (HWED). 1995. *Support Hmong Women Educational Delegation 4th United Nations World Conference in Women, Beijing, China, September 1995*. Washington, DC: Hmong National Development.

Holmes, David, Kate Hughes, and Roberta Julian. 2003. *Australian Sociology: A Changing Society*. Sydney: Pearson Education Australia.

Julian, Roberta. 1997. Invisible subjects and the victimised self: Settlement experiences of refugee women in Australia, in *Gender and Catastrophe*, edited by Ronit Lentin. London: Zed Books, 196–210.

Julian, Roberta. 1998. 'I love driving!' Alternative constructions of Hmong femininity in the West. *Race, Gender and Class*, 5(2), 30–53.

Julian, Roberta. 2003. Transnational identities in the Hmong diaspora, in *Globalization, Culture and Inequality in Asia*, edited by Timothy J. Scrase, Todd J.M. Holden, and Scott Baum. Melbourne: Trans Pacific Press, 119–43.

Julian, Roberta. 2004. 'Living locally, dreaming globally: Transnational cultural imaginings and practices in the Hmong diaspora, in *The Hmong of Australia: Culture and Diaspora*, edited by Nicholas Tapp and Gary Yia Lee. Canberra: Pandanus Books, 25–57.

Julian, Roberta. 2004–05. Hmong transnational identity: The gendering of contested discourses. *Hmong Studies Journal*, 5. 1–23.

Julian, Roberta, Adrian Franklin, and Bruce Felmingham. 1997. *Home from Home: Refugees in Tasmania*. Canberra: Department of Immigration and Multicultural Affairs (DIMA).

Kleinschmidt, Harald. 2003. *People on the Move*. Santa Barbara, CA: Praeger Publishers Inc.

Lee, Gary Yia. 1996. Cultural identity in post-modern society: Reflections on what is a Hmong? *Hmong Studies Journal* [Online], 1(1, Fall):1–14 Available at: http://members.aol.com/hmongstudiesjrnl/HSJv1n1_LeeFr.html [accessed: 27 January 2003].

Lee, Gary Yia. 2004. Culture and settlement: The present situation of the Hmong in Australia, in *The Hmong of Australia: Culture and Diaspora*, edited by Nicholas Tapp and Gary Yia Lee. Canberra: Pandanus, 11–24.

Lemoine, Jacques. 2008. To tell the truth. *Hmong Studies Journal*, 9, 1–29.

Moua, Mai Neng (ed.). 2003a. *Bamboo among the Oaks: Contemporary Writing by Hmong Americans*. St. Paul: Minnesota Historical Society.

Moua, Mai Neng. 2003b. Review of *Bamboo among the Oaks* [Online]. Available at: http://www.mnhs.org/market/mhspress/products/0873514378.html [accessed: 15 January 2004].

Morokvasic, Mirjana. 1988. Cash in hand for the first time: The case of Yugoslav immigrant women in Western Europe, in *International Migration Today, Volume 2 Emerging Issues*, edited by Charles Stahl. Nedlands, Western Australia: Centre for Migration and Development Studies, University of Western Australia, 155–67.

Pfaff, Tim. 1995. *Hmong in America: Journey from a Secret War*. Eau Claire, WI: Chippewa Valley Museum Press.

Pfeifer, Mark E. and Serge Lee. 2004. Hmong population, demographic, socioeconomic, and educational trends in the 2000 census, in *Hmong 2000 Census Publication: Data and Analysis*. Washington, DC and St. Paul, MN: Hmong National Development, Inc. and the Hmong Cultural and Resource Center.

Rex, John. 1995. Ethnic identity and the nation-state: The political sociology of multi-cultural societies. *Social Identities*, 1(1), 21–34.

Richmond, Anthony H. 1988. Sociological theories of international migration: The case of refugees. *Current Sociology*, 36(2), 7–25.

Robertson, Roland. 1992. *Globalisation: Social Theory and Global Culture*. London: Sage Publication.

Schein, Louisa. 1998. Forged transnationality and oppositional cosmopolitanism, in *Transnationalism from Below*, edited by Michael Peter Smith and Luis Eduardo Guarnizo. New Brunswick: Transaction Publishers, 291–313.

Schein, Louisa. 2000. *Minority Rules: The Miao and the Feminine in China's Cultural Politics*. London: Duke University Press.

Schirato, Tony and Jen Webb. 2003. *Understanding Globalization*. London: Sage Publications.

Song, Miri. 2003. *Choosing Ethnic Identity*. Cambridge: Polity Press.

Spivak, Gayatri Chakravorty. 1990. *Post-Colonial Critic*. London: Routledge.

Tapp, Nicholas. 2004. Hmong diaspora in Australia, in *The Hmong of Australia: Culture and Diaspora*, edited by Nicholas Tapp and Gary Yia Lee. Canberra: Pandanus, pp. 59–96.

Tapp, Nicholas and Gary Yia Lee (eds). 2004. *The Hmong of Australia: Culture and Diaspora*. Canberra: Pandanus.

Thao, Cheu. 1982. Hmong migration and leadership in Laos and in the United States, in *The Hmong in the West*, edited by T. Downing and D. Olney. Minneapolis: University of Minnesota, pp. 99–121

Werbner, Pnina. 1998. Diasporic political imaginaries: A sphere of freedom or a sphere of illusions? *Communal/Plural*, 6(1), 11–31.

Xiong, Tou Ger. 1998. *Hmong Means Free*, Video produced by Tou Ger Xiong.

Yang, Dara Carol. 2002. Single … longer. *Hnub Tshiab*, 3(1), 1–2.

Chapter 12
Globalised Cartographies of Being: Literature, Refugees, and the Australian Nation

Tony Simoes da Silva

> Refugees, the human waste of the global frontier-land, are the 'outsiders incarnate', the absolute outsiders, outsiders everywhere and out of place everywhere except in places that are themselves out of place – the 'nowhere places' that appear on the maps used by ordinary humans on their travels.
>
> Zygmunt Bauman (2004b: 80)

This chapter considers the figure of the refugee as the displaced individual through the reading of a number of Australian literary works, which explore displacement 'as an extreme case of a more general modern condition – the powerlessness of the individual caught in the grip of vast collective purposes', to borrow Ian Watt's (1959: 218) comments on World War II prisoners of war. Through a critical reading of selected works aimed both at children and adult readers, I consider the role textual representation can play in creating a different understanding of the subject positions of the mass of individuals arriving on Australian shores. Two main issues are addressed here. First, the role literature can play in the exploration of an identity politics associated with the experience of the refugee as an 'invisible–visible' presence (Benbassa 2008). My concern is the idea of refugee selfhood as a shifting and contingent construct that emerges from relations of production at once historical, political, psychological, and affective. The refugee is in this context a complex site of interaction between past and present, Self and Other, nation and foreign. In my reading of selected literary texts, I trace how this fluid conception of an identity selfhood is both inflected by and in turn then inflects a broader notion of national unity and exclusion. Further, through a detailed textual analysis of selected literary texts, I show how literature can contribute to a fuller and subtler understanding of what it means to be a refugee. The second issue concerns the activist role a number of contemporary Australian writers have sought to play on behalf of refugees and of the very experience of flux associated with displacement. Given the obvious echoes between the writings, the chapter seeks to place these texts in a dialogue with the work of social scientists such as Zygmunt Bauman, Giorgio Agamben, Michel Agier, Peter Nyers, and others.

For Peter Nyers, quoting Randy Lippert, the refugee is 'a new kind of person' (2006: 13), bringing together the complexities of being human in the contemporary times. In the unsettled and unsettling fluidity of the displaced self coalesce some of the issues typical of the contemporary period; hence Zygmunt Bauman's (2004a 2004b) assertion that refugees in particular both embody and create a kind of *zeitgeist* (2004b). Nyers himself uses the term 'refugeeness' to describe 'the various qualities and characteristics that are regularly associated with and assigned to refugee identity' (2006: xv). Stating at the outset his awareness of the term's 'fuzziness' and ambiguity, Nyers nevertheless posits it as a political and as a cultural motif that define and unite the disparate masses of people now on the move from Asia, Africa, and the Middle East towards Europe and the USA, though more likely a shift sideways between nation-states in these regions. In other words, the experiences of refugees are too diverse to be encompassed by a single term. Nyers, moreover, notes that '[t]he politics of being a refugee has as much to do with cultural expectation of certain qualities and behaviours that are demonstrative of 'authentic' refugeeness (e.g., silence, passivity, victimhood) as it does with legal definitions and regulations' (2006: xv). Australia's response to refugees is a case in point, for they are at once perceived as a threat to the nation and as objects of pity; they terrify, but they also unite the hugely diverse body of people that is Australia. In my analysis of the selected texts, I draw on this complex framework to propose a poetics of refugeeness that complements and complicates the politics of refugeeness. *What* is refugeeness? *How* is refugeeness experienced? *When* does refugeeness begin?

Like Nyers, Bauman has stressed the need to understand displacement within complex contextual conditions. Both explain it as the product of a concatenation of factors to which economic rationalism, the work of the International Monetary Fund and the World Bank, and myriad civil conflicts are contributing actively and generously. It is also, increasingly, impelled by climate change that is often felt more acutely in areas and by peoples already experiencing severe hardship. In *Wasted Lives* (2004b), Bauman traces many of these conditions through an analysis of some of the structural frameworks that underpin the morphing of modernity into what he called 'liquid modernity' in an earlier work of that title, more colloquially described as 'globalisation'. Through a focus on the condition of peoples caught up at the intersection between globalism and localism, politics, and economics, Bauman argues that '[t]he production of "human waste", or more correctly wasted humans … is an inevitable outcome of modernisation, and an inseparable accompaniment of modernity' (2004b: 5). Deemed by economic and increasingly political discourses and structures to be 'redundant' (Bauman 2004b: 5), these are people whose existences are apparently superfluous to the needs of global capitalism. Yet, perhaps somewhat perversely, they are also absolutely indispensable to mainstream Western societies; the economies of the great contemporary superpowers such as the USA, China, India, and some of the leading European nations rely on the easy flow of people desperate enough to work for insignificant pay and under appalling conditions. For literary critic Graham Huggan (2007), in words that resonate with

Bauman's own notion of 'wasted lives', these groups are in fact contemporary modernity's 'itinerant underclasses' (p. vii).

One can thus trace in this discussion a clear sense of the condition of 'refugeeness' as permanently fluid and uncontainable. In the work of fiction, can certain metaphors better convey the dynamic nature of refugeeness, which is always, paradoxically, over signified and insufficiently defined? Is the refugee already, 'naturally', associated with certain parts of the world? How might one read the figure of the refugee as a critique of the hegemonic state of world politics post-'9/11', but more broadly the state of world politics in general? A characteristic of both theoretical and imaginative writing is in fact an emphasis on the invisible visibility of the refugee, denied reality (and selfhood) either by material, psychological, or, more often than not, discursive factors. The displaced are 'placeless' and 'stateless', without identification papers, they are 'selfless': identity and selfhood go hand-in-hand. The brutal paradox refugees repeatedly face is that in order to be refugees they need to lose their previous selves; without documentation they are 'outsiders incarnate' – Bauman's take on Agamben's (2005) work. They are also the invisible blood that runs in the ever-more fluid veins of contemporary capitalist structures. As noted, Bauman in particular has now written at length on the role displacement plays in the perpetuation of a capitalist market economy, but this is a view shared by others (Agier 2008, Bauman 2004b, Nyers 2006). Papastergiadis views the impact of the experience of displacement overlapping with that of people who do not move away from their places of residence: 'Countless people are on the move and even those who have not left their homeland are moved by this restless epoch' (2000: 2).

Indeed, as 'a new kind of person' (Lippert), the refugee has become instrumental to a growing strengthening of national discourses. To say that the figure of the refugee and the response it has elicited across the world has highlighted the nation as essentially a product of a complex process of exclusionary inclusion is hardly a novel assertion. In Australia, this is reflected in the constancy of calls for the protection of 'our' borders, and indeed of the 'sanctity of our borders' (Manne 2005, Perera 2009). It is a discourse that has its genesis in the White Ausralia policy, but one with an uncanny ability to make bedfellows of the most disparate individuals. Pauline Hanson is reviled for her language on asylum seekers; but her position was in many ways only a cruder form of those shared by mainstream politicians on both the left and the right. It is a discourse that is especially perplexing in the age of 'borderless market economies'. In *Identity* (2004a), in a series of interviews with the Italian journalist Benedetto Vecchi, Bauman notes that '[w]hether in their consciousness or their subconscious, men and women of our times are haunted by the *spectre of exclusion*' (p. 47, emphasis in the original). What is interesting here is the ambiguity of Bauman's claim, for like Papastergiadis, he refuses to see exclusion as something directed simply at refugees. In an influential piece on alterity, Jacques Derrida (2001) argued in *On Cosmopolitanism and Forgiveness* that that which we fail to do for the other we fail to do for ourselves. That I should summarise it in these terms betrays my own Catholic upbringing but it is perhaps the product of

Derrida's own engagement with Levinas and a Judeo-Christian tradition. This point will be understood later in the chapter with the help of an example.

Working as a literary critic, I am in full agreement with writers such as Bauman, Nyers, Agier, Arundhati Roy, and Vandana Shiva that refugees are the *product* (an intentional use of the market economy vocabulary) of political processes that originate largely in the developed world, but whose consequences are felt primarily elsewhere. The depiction of refugees and their experience in contemporary fictional literature and film is also wide-ranging. Stephen Frears's 2006 *Dirty Pretty Things* is perhaps one such interestingwork, a dense narrative that seeks to combine a strong political edge with genuine empathy for the myriad of peoples living 'invisible–visible' existences in modern-day London. Frears brings into play some of the ideas I have raised earlier by complicating definitions of 'refugeeness', for example, by setting alongside with each other genuine refugees with political exiles and economic migrants. Rather than passing judgement, the film forces viewers to confront the blurred boundaries between any such neat categorisations, and at times denies viewers the outsider position that might enable them to *know* the refugee. Rather than concern itself with 'definitive definitions', to paraphrase Nyers's own discomfort with the term 'refugeeness', *Dirty Pretty Things* foregrounds the materiality of the displaced's existence, at once extraneous to the national body and utterly intrinsic to it. In an exchange in an anonymous car park, Okwe, the Nigerian doctor, a political refugee, is asked: 'How come I've never seen you guys?' He replies: 'We are the people you never see. We wash your cars, clean your houses, suck your cocks.' Okwe's response seems a direct take on Bauman's view on the displaced as the necessary underside of modernity. Refugees provide cheap labour that oils the machinery of capitalism. But the question is posed by a Spanish man, who works as a hotel concierge and runs a sideline business in human organs. As a doctor, Okwe becomes involved when a Somali refugee begins to bleed profusely after 'donating' a kidney in exchange for identity papers. When Okwe asks the man's relatives what the transaction entailed, the answer is straight out of Bauman, Nyers, or Agamben: 'He's English now.' Echoing Bauman, Frears too privileges the blurring of boundaries and ontologies.

In Australia and elsewhere, imaginative writing has emerged as an influential site of representation of the experience of displacement and of the contextual frameworks that explain them. Literature, especially though not exclusively when addressed to children, draws on its 'shaping force' (Boehmer 2007: 6) to counter the destructive impact of political discourses on refugees and the displaced more generally. It has also gained importance as a tool of political and social activism on behalf of refugees and their condition. As David Attridge suggests, '[t]he singularity of the literary work is produced not just by its difference from all other works, but by the new possibilities for thought and feeling it opens up in its creative transformation of familiar norms and habits: singularity is thus inseparable from *inventiveness*' (2004: 11). Although there is a risk that some acts of representation of the refugee experience will echo Karl Marx's famous dictum 'They cannot represent themselves; they must be represented' (Said 1978), literary works view refugees as 'outsiders incarnate' (Bauman 2004b: 80) in need of a voice. Australian writers such as Blanche D'Alpuget, John Marsden, Matt Ottley, Eva Sallis, Alice Pung, Thomas Keneally, and countless others have sought to use the affective power of literature Attridge refers to as a means to counter 'the demonization of asylum seekers', which Suvindrini Perera notes is 'once again a popular sport' (2009).

The persistence of this view can be seen in political scientist Robert Manne's *Left Right Left: Political Essays 1977–2005* (2005). In it, Manne traces Australia's response to refugees to the nation's relatively benign coming into being. He wrote that 'the atmosphere … reflected the Australian incapacity to understand (given the tranquillity of our political history) the kind of political conditions that make people flee from regimes, or indeed the physical deprivations and moral humiliations that people living in squalid refugee camps undergo' (p. 385). Manne himself would eventually change his opinion of Australia's tranquil political setting, notably after the publication of the report on the Stolen Generations. In a surprisingly candid and perplexing admission by one of Australia's best-known political commentators, Manne confessed in the aftermath of the publication of the *Bringing Them Home* report on the Stolen Generations that he had known next to nothing about the treatment of Indigenous Australians. This despite the fact that he was in the *business of knowing* about Australia's past and present. While being tempted to be cynical about such political *naïvete*, it is worth placing it within a white, settler Australia psychosis that defends the fear of invasion and the 'right to secure our borders' perhaps precisely because as a modern nation, Australia is the result of invasion and porous borders. Not uncommonly do we fear the open doors through which we ourselves walked, however sedulously.

Manne's position is useful here because it captures the essence of Australia's historical unease about refugees and the so-called 'illegal arrivals'. Ironically, although Australia in its modern form is the product of illegal invasion, each successive wave of migrants or asylum seekers since World War II has created its own micro climate of hysteria. In an essay entitled 'Indo-Chinese Refugees', concerned with the Australian response to the arrival of a number of boats bearing Vietnamese refugees between 1976 and 1978, Manne noted that 'in the space of two weeks a predictable arrival on our shores of a small number of refugees had activated our deepest collective neurosis' (2005: 384). The 'invaders' then were 'Indo-Chinese Refugees', but they might equally have been 'Middle Eastern Refugees', the title of a later piece he wrote in *The Sydney Morning Herald* in 2001 (Manne 2005). Uncannily, of course, both pieces might easily be published today with little or no change, either in content or tone. The Australian 'collective neurosis' remains as alive in 2012 as it did in 1976, 1978, 1999, and 2001. In 1976–78, Manne described the reactions of iconic Australian politicians such as Bob Hawke and Gough Whitlam to the refugee arrivals as 'suggestive of [men] (and a culture) which have no grasp of the ruthlessness of the political process in the twentieth century' (Manne 2005: 385). There is a certain irony in the fact that both politicians were on the left in Australian politics and within the Great

Australian Myth of the nation; they were, thus, supposedly on the side of the underdog.

By the time he wrote 'Middle Eastern Refugees' in 2001, Manne was well on the way to becoming one of the strongest voices speaking on behalf of refugees and all manner of illegal arrivals. Once again, the irony is that as if to prove Manne's point about Australia's inability to grasp the complexity of international politics and postcolonial history, the piece he wrote on the new arrivals now fleeing 'fundamentalist Afghanistan and Saddam Hussein's Iraq)' (2005: 386) would not have been out of place in 2012 He wrote then:

> As always in politics, the government's language helped shape public perceptions. By describing the illegal operations as 'people smuggling' rather than as a trade in human misery, it made the refugees appear not so much as wretched human beings but as some new sinister species of contraband. By describing the refugees as 'queue-jumpers', it made them appear as ruthless egoists, selfishly satisfying their own needs at others' expense. In fact, in the arbitrary world of the refugee camp, there does not exist anything so civilised as a queue. By systematic use of language of such kind, the Howard government destroyed whatever slender possibility existed for popular understanding of the refugees' plight. (Manne 2005: 387)

I quote at length here for two main reasons. On the one hand, had Manne substituted 'Hawke' for 'Rudd', he might equally have been describing the response of Australia's former Prime Minister (2007–10) to the ongoing arrivals of boats carrying refugees fleeing Sri Lanka, Iraq, or Afghanistan. On the other hand, it is perhaps much too simplistic to accuse Australian politicians of stoking a lack of understanding of and consequently of empathy with the refugees' situation as if Australians were unable to make up their own minds. To put differently, the politicians' response is actually fed by a general Australian insularity that sees even former refugees turn against the new arrivals. As Manne again would note, in a longer disquisition on Australia's treatment of refugees in *Sending Them Home* (2004), Australia's poor grasp of world politics has led to a tendency to think that 'worthy refugees waited patiently in a Third World Country [while u]nworthy refugees used their money to engage the services of people smugglers. Willingness to pay people smuggles stripped unworthy refugees of any claim to sympathy. It represented in them a kind of indelible moral taint' (2004 398). The latter perception is particularly galling for Manne because, as a Jewish Australian, he knows that many Holocaust survivors too had to resort to the services of people smugglers. Manne's views also echo Wells's view that '[w]ithout discussing [the] contextual background – the conditions that provoke refugees – it is wistfully misleading to presuppose that 'orderly queuing' and 'following the correct channels' are real options facing those fleeing repression or social chaos' (Wells 2005: 15).

Manne's essays and his own shifting positionality highlight the rift that was beginning to emerge within Australian society about the nation's response to and treatment of refugees and the role of literature in this setting. Blanche D'Alpuget's controversial depiction of the brutalisation of Indo-Chinese refugees on Malaysian shores in *Turtle Beach* (1981) was one of the earliest literary engagements with the experience of refugees, and at the time it led to a protracted diplomatic rift with Malaysia. Like D'Alpuget, more recent authors such as Marsden, Ottley, Sallis, Shaun Tan, and Linda Jaivin have also used their writing as a means to engage in the kind of debate thus outlined. In the case of Marsden and Tan, this is directed especially at children, and in this way their work seeks both to change the perception Australians have of refugees and set in place the conditions for a new 'national mindset' by educating the generation who will rule in the future. The following paragraphs briefly examine a number of these works to show how they represent the idea of a 'refugee consciousness' and complicate the very experience of 'refugeeness'.

John Marsden and Matt Ottley's work, *Home and Away* (2008) tells the story of a family whose mundane existence is turned upside down by civil conflict. Alternating between illustrations with the quality of photographs and a range of childish drawings of varying levels of ability, the book sets out to create an intimate bond between the family and readers by stressing the ordinariness of the family's daily life. Photos are also still, despite the power of Photoshop, a powerful repository of familial and personal memory, all the more effective for the way in which they always inevitably slip into nostalgia. Opening with a brief portrait of the family which the reader 'accesses' by directing the mouse cursor to key actors such as 'Mum', 'Dad', 'Me', 'Claire', and 'Toby', *Home and Away* allows readers to see themselves reflected on the page, as it were. These are people 'like us' and indeed the point of the book is that these people are 'us'. That the characters are 'white' and blue-eyed is central to the work's ability to connect with the mythical 'ordinary Australians'; it matters little that the typical Australian today is unlikely to have blond hair and blue eyes, and Marsden knows it well. However, he knows also that as a society Australia is regulated by a raft of normative discourses to which whiteness remains core (Hage 1998, Rutherford 2000). In this way, *Home and Away* avoids making the would-be-refugees non-white to show that they are 'just normal people' whose humanity Australian readers share. The family's likes and dislikes, their dreams and fears all are revealed in the first three pages in a way that lends pathos and bathos to the experiences soon to befall them. These are an ordinary, 'everyday family'. As with race, it is impossible to overlook the heteronormative function the family plays in this narrative – suburban family of Dad, Mum, and three children. When the war starts, on 'April 27' (Marsden and Ottley 2008: 5), the illustration is of an aquarium coloured with bright red splashes and a child-like cowboy on the family's TV screen. The impact of the disruption of the family's most intimate space is underlined by the fact the sign on the TV screen is both one of the most natural and playful and an insidious reminder of the abrupt and unexpected arrival of civil conflict that will disrupt their daily routines. Up-to-now there have been sufficient details, visual and linguistic, to suggest that the setting of the story is Australia, but there is also a clear attempt at playing with the suspension of disbelief typical of fiction. To put it differently, the book provokes readers into 'disbelieving' that war will ever occur in Australia, and that it might make refugees of perfectly ordinary, even dull people. In a globalising world where American popular culture dominates, the family's portrait positions them as urban, modern, contemporary, but crucially 'a-national'. They are a modern 'world family' and the setting of the story should be Sydney, Australia or Copenhagen, Denmark, Toronto, Canada, or Manchester, UK. Although readers in Australia will be tempted to identify the family as Australian, this is essentially the product of the complex affective networks fiction unleashes and that it seeks to engender more generally. In this way, the book both harvests and neutralises the power of surprise, for as readers, in Australia one is naturally – that is to say, culturally – programmed to assume that war and turmoil break out always elsewhere, away from home. To an Australian audience, the title of the book, *Home and Away*, is itself part of this elaborate process of interpellation of the Australia reader. As the title of one of the longest running and best-known Australian TV series, the iconic 'soapie' 'Home and Away' signifies Australia as essentially beautiful, plentiful, and trouble-free. The TV series glamorises and glorifies the thin veneer of an Australian ordinariness that makes us lovable like every-one-else templates, whose symbolic value Marsden and Ottley resignify as a means of foregrounding the humanity of the refugees arriving on Australian shores. In other words, they underline the tension between *our* view of the world and *their* view of the world, but simultaneously un-hinge the meaning of the possessive pronouns. In their book, directed at children, Marsden and Ottley reverse this idea by bringing the war home, as it were. Thus, they allude also to Australia's own role in the recent wars in Iraq and Afghanistan, but even earlier in Vietnam. Importantly, and to reiterate, the book does not obfuscate its Australian setting, but reveals it carefully, strategically, and provocatively. The slow unveiling of the vulnerability of Australia and Australians plays directly into the critique the work undertakes of the country's unwillingness to grasp the vulnerability of those seeking refugee on its shores. In John Marsden and Matt Ottley's *Home and Away*, fear plays a central role, but rather than the conventional tool that demonises refugees, it seeks to humanise them. The authors draw on the figure on the refugee not as a foreign Other, but as the known self; we, whoever this 'we' might be, are all refugees *in potentia*.

Like *Home and Away*, Eva Sallis's *The Marsh Birds* (2005) takes as its main theme the situation faced by refugees, focusing specifically on what Manne identifies as 'Middle Eastern refugees'. The story of Durgham, a 12-year-old Iraqi child trapped in Damascus, Syria after his parents fail to show up at the agreed meeting point, allows the novel 'to theorise' the coming into being of the refugee as a protracted and painful process. Taking the reader into a child's mind, it seeks to imagine what it is like to be alone, afraid, and away from home. The use of a child narrator, one of the most common ways of representing individuation, allows Sallis to stress the lasting damage of the experiences Dhurgham will undergo in Damascus. The novel opens with the following lines: 'Dhurgham's chest hurt. His head hammered in its bed of pain. It was the third day and all his limbs ached' (Sallis

2005: 3). Although narrated in third person, the novel takes the reader into the kind of intimate experience that only the personal self can access. Indeed, the detached third-person narrative voice may also be interpreted as a coping mechanism for the abandoned and soon-to-be abused child, but most of all it foregrounds the vulnerable child. In the kind of portrait that brings into play the complexity of 'refugeeness', Dhurgham is both a lonely and terrified young boy and the child of a wealthy Iraqi family who fled carrying with them substantial amounts of money. Although a victim, he is a product of privilege; both facets frame the unstable ontology of refugeeness in the novel. Sallis's depiction of Dhurgham's character, particularly the abuse he endures at the hands of a paedophile in Damascus, a man who also robs him of most of his money, prevents the novel from appearing to presume that the brutalised refugee is born only when he or she leaves Indonesia or Malaysia for Australia. Rather, *The Marsh Birds* posits refugeeness as a staged process, which, for Durgham and his family, begins when he flees Baghdad, and one which for him begins to develop in Damascus, while for his parents it terminates there. When Dhurgham eventually buys his way to Indonesia through the services of people smugglers, he resumes the making of his 'refugee' self, who will then linger in a transit camp on an Indonesian island. The irony the novel identifies is that in this space, Dhurgham's refugee self now fades away. Unless housed within a UNHCR-recognised camp, where he will undergo the full force of the discursive apparatus of refugeeness, Dhurgham and others like him are perceived as 'illegal migrants', 'aliens', or potential 'terrorists'. Deprived of the right conditions for its existence, Dhurgham's refugeeness goes into abeyance, as it were. For as Bauman, Agamben, Sivanandan, and others stress, the experience of displacement is best understood as a synonym for the inherent contingency that obtains in definitions of 'migrant', 'illegal alien', 'economic migrant', and 'asylum seeker', for example. These, after all, are not innocent categories; they are produced by legal and political discourses deployed by nation-states intent on controlling the movement of the roaming multitudes, whose sheer number defines as the enemy. By privileging in her book the spectrum of more or less identifiable stages in the making of a refugee, *The Marsh Birds* provides the means for a richer understanding of the trauma refugees often take forth into their new lives.

As an Australian, Sallis has long been involved in campaigns on behalf of refugees, especially those held in detention centres. Sallis's book is part of this concern with human rights that brings her writing into the sphere of political activism. One of the most remarkable aspects of *The Marsh Birds* is its ability to stand up as a work of fiction of considerable power, while not holding back on its politics. Didactic in ways that Marsden and Ottley also deployed, the novel sets up a stage for a semi-rhetorical conversation between the detainees held in camps in Australia and an Australian population, who, for the most part, has no understanding of the meaning of refugeeness or the conditions for its existence. In *The Marsh Birds*, as noted earlier, politics is reflected also in the fact that Australia and Australians have been instrumental in the creation of the present Iraqi conflict. As one of the detainees, Aziz, ponders at one point: 'Why do Australians do this to us? We are not criminals' (2005: 114). In a twist that is characteristic of Australian novels and children books concerned with refugees and their experience, Sallis gives us a Dhurgham who is able to distinguish between 'bad, uncaring Australians' and 'good, caring Australians'. In this way, *The Marsh Birds* echoes Robert Manne's views on 'worthy' and 'unworthy' refugees. All in different ways stress the importance of not generalising and the potential for empathy shared by all human beings outside of social markers of race, gender, and national identity. When in *The Marsh Birds* Mrs Azadeh, trapped in a detention centre, remarks that 'No one in Australia know about what happens here. Until you are freed you are not in Australia' (2005: 115), she underlines the fair-minded nature of most Australians; as she suggests, only lack of awareness prevents Australians from intervening to stop the incarceration and dehumanisation of refugees like her.

Yet as the reaction of the Labor Governments under Kevin Rudd and Julia Gillard has shown, Sallis's work and that of many other creative writers overestimates the power of the word to change the world. The questions posed in the blurb of Sallis's book are central to this issue: 'What do you do when you belong nowhere, with no family, no homeland, and no hope for the future? Who do you become?' Ultimately, the work of imagination will only ever be able to gesture towards an act of empathy with the Other, but perhaps that is all it can aim and hope to achieve. Besides, as Edward Said notes when he quotes Karl Marx's condescending views, there is always a risk that what the writer imagines to be the world of the refugee, the placeless refugee with no family and perhaps without identity papers, is itself a flawed portrait of the Other. It is in this way an imposition where the act of giving – of the self to the Other through the act of empathetic sharing – returns as a variation of (neo)colonial Othering. 'Who can write as a refugee?' is a question worth asking not to negate the ability of writers such as Marsden and Ottley and Sallis to act on behalf of refugees, but to ensure that one is aware of the benefits of benevolence. To do good to an Other is to do good to oneself, to return to the Judeo-Christian point raised in the beginning of the chapter. This may be a cynical and hopeless question, but it is one that needs to be asked. The answer, at least a kind of answer, will necessarily involve coming to grips with the intentions of the writer, his or her ethical and political credentials.

One of the most commonly accepted view on refugees today is that they constitute a 'problem'; that they contribute to a 'crisis' and that as group of people they are caught up in what is referred to, with more or less genuine sincerity, as a 'plight'. Andrew Wells (2005: 15) has suggested that refugees show up and in turn are themselves part of a 'globalisation of the powerless'. They constitute a fluid mass of people roaming the planet in search of a place to live a relatively safe and ordinary existence. Echoing the work of social scientists and political activists, the historian Wells argues that globalisation has created the refugee problem as is known today, notably through the pressure which organisations such as the International Monetary Fund and the World Bank have imposed on fragile nation-states, whose survival has been seriously compromised by the policies they have been forced to implement in order to belong to a 'community of nations', a new world order of sorts. As seen earlier in the chapter, Bauman (2004b) goes so far as to argue that the production of a 'wasted people' is indispensable to the ongoing resilience of capitalist power structures. And the pace at which that process is occurring in the West is nowhere as destructive as the impact its counterpart has had in the 'developing world'. But the 'production' of refugees is a far more complex issue than both Bauman and Wells allow for. For the social scientist and the historian, driven by an empirical and factual impulse, refugeeness is best understood with reference to well-established and researched points of origin. In contrast, it is proposed here, imaginative literature is freer to create a space for acts of empathy with characters who might not be real people, but who are *like* real people, potentially 'like us'. However obvious this may seem, the gap between the imagined refugee and the real refugee has the ability to bring into being fleeting insights into the being of a brutalised and dehumanised Other that fact and empiricism cannot convey.

References

Agamben, Giorgio. 2005. *States of Exception*. Chicago: University of Chicago Press.

Agier, Michel. 2008. *On the Margins of the World: The Refugee Experience Today*. Cambridge, UK: Polity Press.

Attridge, David. 2004. *J.M. Coetzee and the Ethics of Reading: Literature in the Event*. Chicago: University of Chicago Press.

Bauman, Zygmunt. 2004a. *Identity: Conversations with Benedetto Vecchi*. Cambridge, UK: Polity Press; Malden, MA: Blackwell Publisher.

Bauman, Zygmunt. 2004b. *Wasted Lives: Modernity and Its Outcasts*. Oxford: Polity Press.

Bauman, Zygmunt. 2008. *The Art of Life*. Cambridge, UK: Polity Press.

Benbassa, Esther. 2008. Préface, in *L'histoire des minorités est-elle une histoire marginale?* Edited by Stéphanie Laithier and Vincent Vilmain. Paris: PUPS, Presses de l'Université Paris-Sorbonne, 7–10.

Boehmer, Elleke. 2007. Postcolonial writing and terror. *Wasafiri*, (51, Summer), 4–7.

D'Alpuget, Blanche. 1981. *Turtle Beach*. Ringwood, Victoria: Penguin.

Derrida, Jacques. 2001. *On Cosmopolitanism and Forgiveness*. Translated by Mark Dooley and Michael HughesPublisher: Routledge. London: Routledge.

Hage, Ghassan. 1998. *White Nation: Fantasies of White Supremacy in a Multicultural Society*. New York, Annandale, NSW: Routledge.

Huggan, Graham. 2007. *Australian Literature: Postcolonialism, Racism, Transnationalism*. New York: Oxford University Press.

Jaivin, Linda. 2006. *The Infernal Terrorist*. London: HarperCollins.

Keneally, Thomas. 2004. *The Tyrant's Novel*. London: Sceptre.

Lippertt, Randy. 1999. Governing Refugees: The Relevance of Governmentality to Understanding the International Refugee Regime. *Alternatives*, 24.3. 295–328.

Manne, Robert. 2004. *Quarterly Essay 13: Sending Them Home: Refugees and the New Politics of Indifference*. Melbourne, Vic.: Black Inc., 2004.

Manne, Robert. 2005. *Left Right Left: Political Essays 1977–2005*. Melbourne: Black Inc.

Marsden, John and Ottley, Matt. 2008. *Home and Away*. Sydney, NSW: Hachette Livre Australia.

Nyers, Peter. 2006. *Rethinking Refugees: Beyond States of Emergency*. London: Routledge.

Papastergiadis, Nikos. 2000. *The Turbulence of Migration: Globalization, Deterritorialization, and Hybridity*. Malden, MA: Polity Press.

Perera, Suvindrini. 2009. What gain in stopping the boats? *The Sydney Morning Herald* [Online] Available at: http://www.smh.com.au/opinion/politics/what-gain-in-stopping-the-boats-20091111-i9y1.html [accessed: 13 November 2009].

Pung, Alice. 2006. *Unpolished Gem*. Melbourne: Black Inc.

Rutherford, Jennifer. 2000. *The Gauche Intruder: Freud, Lacan and the White Australian Fantasy*. Carlton, Vic.: Melbourne University Press.

Roy, Arundhati. 2004. *The Ordinary Person's Guide to Empire*. London: Flamingo.

Said, Edward. 1978. *Orientalism*. London: Routledge.

Sallis, Eva. 2005. *The Marsh Birds*. Crows Nest, NSW: Allen & Unwin.

Sivanandan, A. 1990. *Communities of Resistance: Writings on Black Struggles for Socialism*. London and New York: Verso.

Shiva, Vandana. 2005. *Earth Democracy: Justice, Sustainability and Peace*. Cambridge, MA: South End Press.

Tan, Shaun. 2006. *The Arrival*. South Melbourne: Lothian Books.

Watt, Ian. 1959. 'Bridges over the Kwai', *The Listener*, 6 August.

Wells, Andrew. 2005. 'Globalisation of the Powerless: A Zone of Instability and the Disabled State'. In *Seeking Refuge: Asylum Seekers and Politics in a Globalising World*, Edited by Jo Coghlan, John Minns and Andrew Wells. Wollongong: Wollongong University Press. 15-25

Wilson, Ronald. 1997. *Bringing Them Home: Report of the National Inquiry into the Separation of Aboriginal and Torres Strait Islander Children from their Families*. Sydney: Human Rights and Equal Opportunity Commission.

Chapter 13

Dispossession, Human Security, and Undocumented Migration: Narrative Accounts of Afghani and Sri Lankan Tamil Asylum Seekers

Ruchira Ganguly-Scrase and Lynnaire Sheridan

Introduction

In the globalised world of the twenty-first century, material and symbolic goods travel relatively freely across national borders. At the same time, movements of people, or at least particular categories of people, are becoming increasingly understood as a problem in need of control (Briskman and Cemlyn 2005; de Haas 2007; Turner 2010). Migration has become 'one of the most controversial areas of policy and practice facing virtually all countries' (Crawley 2006: 25). Perceptions of porous boundaries and unlimited opportunities coexist in the public imaginary with hardened attitudes towards desperate humans who seek to cross-national borders without authorisation by receiving states. Throughout the Global North, humanitarian ideals of social justice towards asylum seekers have given way to a preoccupation with national security and border control (Ganguly–Scrase et. al 2006; Innes 2010; Porter 2003; Sales 2002; Stalker 2001) and the consequent criminalisation of 'undocumented' migrants who arrive without authorisation (De Giorgi 2010; Fekete and Webber 2010; Hörnqvist 2004; Welch and Schuster 2005). When formalised in law, these take the form of increasingly restrictive immigration legislation and regulations, intensified border controls, carrier sanctions, deterrent policies, and return migration policies (de Haas 2007: 823–24). While Fassin and d'Halluin (2007: 308) point out that 'there has never been a Golden Age for refugees' and that '[a]lways, in practice, asylum has come second to nations' economies and securities', the hardening of global attitudes towards asylum seekers is reaching unprecedented levels in popular and institutional discourses. Even previously generous states, such as Denmark, are seeking to reduce their intake of asylum seekers (Betts 2003).

Asylum seekers embody 'the contradiction between the national logic of migration control and the transnational logic of international migration' (Castles 2007: 31), and are, thus, disruptive to visions of a 'national order of things' (Blommaert 2009; Limbu 2009; Turner 2004). Tazreiter (2008) argues that '[t]he arrival of persons without a legal right to stay in a nation-state is often seen as a challenge to the sovereignty of that state'. As a consequence, governments of receiving states strive to contain and repel those whose life trajectories belie the salience of nation states under conditions of globalisation. In defiance of international treaties to which they themselves have been signatories, governments around the world are adopting legislation designed to restrict citizenship, secure national borders, and send a strong message of deterrence to those contemplating unauthorised migration. These include stressful assessment procedures, detention under inhumane conditions, denial of the opportunity to work and to participate in civic life, and barriers to family reunion, social security support, and a range of health services (Rees and Silove 2006).

The association of unauthorised travel with securitisation discourses has given rise to suspicion towards asylum seekers. Once a legitimate political status, the identity of 'asylum seeker' has been re-classified as 'deviant' and is now highly stigmatised, being 'synonymous with "sponger", "beggar", "cheat", and "scrounger"' (Linden 2007: 123). Advocates for the rights of asylum seekers are labelled 'naïve', or 'politically correct', or otherwise marginalised (Fairclough 2003; Poynting and Mason 2007). In the receiving countries of the Global North, asylum seekers are represented as an 'out-of-control, agentless, unwanted natural disaster' (Wodak et al. 2008: 287). A common device employed by politicians and journalists alike to convey images of non-personhood is the liquid metaphor (Bleasdale 2008), which discursively transforms asylum seekers into a mindless, overwhelming, and potentially unstoppable mass. Within the contrasting debates between the undesirability of undocumented migrants and advocates of the humanitarian intake of asylum seekers, rarely are the perspectives of those seeking refuge taken into consideration. Based on research among internally displaced people in Afghanistan and Sri Lanka, this chapter examines the intentions of those seeking refuge.

The Significance of Asylum Seeker Perspectives

There has been scant research into the agency of potential asylum seekers, that is, their intentions and the structural enablers and constraints they encounter in fulfilling those intentions, prior to setting out for a host country. More typically, and often with benign intent, public representations serve to obscure the intentionality of asylum seekers. Academic discourses, for example, tend to focus on the macro-political causes of refugee flight. While essential for understanding the wider context of asylum seeking, such approaches tend to frame refugees collectively as reactive 'masses' or 'surges', which 'spill helplessly across the globe away from war, famine, and persecution in search of sanctuary' (Shawcross 1979: 3). Rose (1981: 8) depicts refugees as

> … human flotsam and jetsam caught in the cross-currents of conflicts which are not of their direct concern, … untargeted victims, bystanders sucked into the maelstrom then washed ashore (or along a muddy trail or fetid campsite) with other frightened hungry and bewildered displaced persons.

In rejecting the use of the term 'dynamics' in relation to the movement of refugees, Kunz (1973: 131) argues that

> … when used in social sciences [the term] suggests the existence of an inner self-propelling force. In the writer's view this inner force is singularly absent from the movement of refugees. Their progress more often than not resembles the movement of the billiard ball: devoid of inner direction their path is governed by the kinetic factors of inertia, friction and the vectors of out-side forces applied on them.

In such discourses, the intentionality of the individuals caught up in such 'surges' is less salient than the actions of states or antagonistic ethnic groups.

In their focus on the suffering and trauma experienced by refugees, humanitarian discourses can also homogenise the refugee experience, drawing on powerful categorisations such as 'women and children' in order to engage the sympathies of ever more sceptical Western donors (see Malkki 1995). Moreover, the 'malignant positioning' (Sabat 2003) of asylum seekers in popular and political discourses is an issue of great concern, since, rather than overlooking or subordinating the intentionality of asylum seekers, it instead misrepresents and misattributes that intentionality in terms of greed and wilful deception. For Sinapi (2008: 534) the 'connotation of the term "refugee" has changed .... increasingly … [being] suspected of secrecy and lies, of being a "false refugee"'. The aim of our research is to challenge these assumptions and demonstrate the complex decision-making processes on the ground.

Attention to the intentionality of individual asylum seekers can forestall dangerous 'category errors' (Clarkson 2003), such as merging the category of 'people' with 'water', which are born of confusion, misinformation, and ideological contamination. An Australian study by Pedersen et al. (2006) connects the kinds of negative beliefs about asylum seekers outlined above with false beliefs on the parts of government officials and the public, and proposes that the correction of those beliefs might offer a way forward to more equitable treatment of this vulnerable category of migrant. This position is supported by Pearce and Stockdale (2009), who quote one British participant, who had expressed particularly negative views on asylum seekers by saying '[i]f I could understand why they left their country and stuff and I knew them I might have a different view'. Another respondent in the same study had commented 'I can't see what they have to offer us, but maybe if I was educated then I may have a better understanding of it' (2009: 152).

In order to comprehend the complexities of asylum seeking, there is a pressing need to prioritise the voices, experiences, expectations, and explanations of those at the centre of the debate—asylum seekers themselves (Dwyer 2008; Zimmerman 2009). Attending to the voices of asylum seekers can then serve to challenge the dominant perspectives that inform the creation of policy (Zimmerman 2009: 204). When the voices of asylum seekers are rendered audible through grounded qualitative research, the stories they tell reveal intentions and circumstances that differ markedly from the discrediting representations outlined above. Moreover, they confirm asylum seekers as active agents rather than passive and helpless victims, who, as noted by Moore and Shellman (2007: 812), are 'making choices under highly constrained circumstances'.

As McKee (2003) observes, how people make sense of their own life experiences can be important in determining their survival. However, despite the centrality of motivations in the legal and normative classification of refugees, plans and intentions do not ensure safe passage to a peaceful life, as the high rates of asylum seeker mortality attest. A concrete example of thwarted intentionality is provided in Antonopoulos and Winterdyk's (2006) account of a dead Kurdish refugee, who was found to have Greek drachmas, Italian liras, German marks, and American dollars sewn on to his belt—a clear indication of the route he had hoped to follow. Like other forms of migration, the movement of refugees is shaped by interactions between people, their resources, and their structural contexts (Lindley 2010). By seeking and recounting the lived experiences of prospective asylum seekers, a nuanced picture emerges that challenges a number of preconceived ideas.

Methodology

As Australian researchers, we purposefully chose to focus on the experiences of the internally displaced in Afghanistan and Sri Lanka as the Australian government has identified these two countries as having the push factors that encourage people to seek asylum (AAP General News Wire 2009). Subsequently, these communities are of interest to both Australian and international organisations as decisions made by the Afghani and Sri Lankan refugees will impact on refugee-receiving nations for, at least, the short to medium term. Yet, in public discourse, they are the most maligned despite these structural factors that currently compel people to flee their homeland.

In general, research on decision-making by asylum seekers is based on retrospective interviews with refugees. Agier (2002: 363) stresses the importance of understanding the precise context of refugees' situations given the diversity of the places, histories, and trajectories of their lives. Interviewing in the place and time where decisions are being formulated by potential asylum seekers enables the researcher to recognise temporal and spatial aspects of decision-making, issues, and contexts specific to particular groups and emergent patterns in data transcripts. Perceptions of time within the context of particular lives affect larger macropolitical processes, by shaping the assessment and intentions of social actors. When potential asylum seekers are considering flight options, their temporal orientation is towards the future. Brown and Michael (2003: 4) point to a 'need for scholarship to engage with the future as an analytical object, and not simply a neutral temporal space into which objective expectations can be projected'. After resettlement, 'migrants may be far removed in both time and space from their experiences of departure so that their reasons for leaving no longer have the relevance that they once did' (Collyer 2010: 279). While studies undertaken after settlement have provided useful data, asylum claimants are likely to emphasise reasons that they believe decision-makers want to hear when arguing their case after arrival in the host country (see Barsky 1995). Koser and Pinkerton (2002) claim that people who have reached a country of potential asylum may be inclined to focus on positive experiences or to misrepresent their experiences. By contrast, the present study focuses on those who are potential asylum seekers. To understand the complex decision-making processes of internally displaced people intending to travel via undocumented means to seek asylum, a qualitative methodological framework comprised largely of semi-structured in-depth interviews was carried out with people in their countries of origin.

Our approach was to let the voices of the displaced dominate because they are the ones who make the decision and then experience the risk of unauthorised migration across borders. This is particularly important as the undocumented migrants have little opportunity to express their views to the broader society. Becker (cited in Taylor and Bogdan 1998) suggests that if a researcher must choose to present 'reality' from someone's point of view, why not choose the powerless who have few avenues for exposing their views. Moreover, Strauss and Corbin (1994) state that it is crucial to present the perspectives and voices of the group studied rather than the assumptions made by others.

Indeed, previous research by Sheridan (2009) on the unauthorised migration of Mexicans to the United States has revealed that while some actively sought to avoid risk by seeking out pseudo legal means by utilising documents belonging to others in order to cross the border entry checkpoints safely with legal documentation, others relied on undocumented migration in the rugged terrain. Interestingly where there was a perception of an increasing level of risk in the homeland, there seemed to be a greater propensity for undocumented travel if the reward would be a chance at a safer and more prosperous life in the longer term. It is, therefore, not clear why asylum seekers would select riskier undocumented journeys if safer options were available. Moreover, the complexities of decision-making is such that asylum seekers may not always seek to settle in countries of the industrialised Global North; instead, they may prefer to live among familiar cultures, bordering their own nation states which have currently reviled and dispossessed them.

Findings

Interviews were carried out over a period of several months in 2010–2011 among Sri Lankan Tamils in Jaffna and Afghanis in Kabul and Jalalabad; the latter were drawn from several different minority ethnic groups, namely, Baluchis, Hazaras, and Tajiks.

Our findings show that most people fled to neighbouring countries; over the protracted period of conflict, most Afghanis had been in Pakistan and Sri Lankans were living in the southern Indian state of Tamil Nadu—only to return in periods of relative calm. While all Sri Lankan Tamils interviewed were planning to undertake unauthorised travel to a Western country to seek asylum, a number of Afghani respondents expressed the desire to settle in nearby countries that they had previously travelled to. This difference is in part due to the diverse socio-economic and cultural backgrounds of the Afghanis, ranging from wealthy businessmen, who had fallen on hard times, to professionals to petty traders. In comparison, Sri Lankan Tamils were from much poorer, working class backgrounds; many also had relatives in Western countries and, thus, were prepared to risk the journey. Although both societies have been affected by civil wars resulting in displacement, the causes and manifestations of conflicts are distinctive. Therefore, we examine the responses of our participants separately. However, we bring together their perceptions of the role of international organisations, particularly that of the United Nations (UN) agencies in assisting the displaced since there is a commonality in their experiences.

Afghani Experiences of Displacement

Increases in violent encounters at home were the primary motivation for participants to relocate to another, relatively safer, part of Afghanistan. Living in a country with many different cultural groups and differing political affiliations, participants found circumstances in their communities changed quickly, as well as frequently, during conflict. People fled when their cultural identity or past affiliations meant that they were perceived to be allied with the 'then' enemy of that community. The enemy, rather than being one group consistently, shifted and changed with different facets of the war and the dominance of one or other cultural group in a region.

A 32-year-old female college professor, from a Tajik community fled when her father was murdered for unknown motives. A Hazara man in his late 40s was forced out of his community when he was suspected of being a spy for the Taliban; in his words: 'First my father was killed at the hands of Mujahideen and then my nephew was killed by Taliban. We would have met the same fate if we did not migrate.'

On the other side of the conflict, a middle-aged Baluchi male was harassed by the Taliban:

> They threatened to blast my shop as well as target my family. Honestly, I have no political affiliation with any party or any connection with army. One day, they fired in the air in front of our house which frightened us a lot. We had only

> one option to migrate to some safer place. I left my home along with my elderly mother, three kids, and a wife. Now I am living in extreme poverty.

A nurse in her early 30s also suffered a Taliban attack: 'Many girls including me were made victims of sexual harassment that night. It was very hard to bear such humiliation and mental torture.' Businessman Mr Askari was at risk as running his business required political alliances and the family became increasingly unpopular among different groups. He added 'in the last days of 2002, my brother was killed.… It was the time of high uncertainty for us and we had no idea of our destination.'

A 52-year-old poetry-loving Tajik male, appropriately named Omar Khyayam, was living in Gardez district, a relatively peaceful area until the Taliban stepped up attacks. Another Tajik male, Dr Ahmed, from Surkhi Parsa district also a relatively peaceful region, had left just before crisis point:

> [W]hen we were leaving, Taliban had almost reached our area. Afterward, our people also took up arms against Taliban because there was no other option left except to face them. One of my relatives has lost their two sons at the time of their migration.

It goes without saying that the displacement experiences were chaotic and traumatic. The decision to leave was quick and people only left at the point they feared for their lives. In general, in the case of the Afghanis, there was no migration plan; instead, they were fleeing without a clear destination or fleeing to family or trusted friends.

A woman informant explained that she escaped without considering a destination or help:

> It was not possible to ask for anybody's help. The whole area was engulfed in a war. Saving one's own life than [that of] others was the top priority of every one at that time. We neither had time to ask for somebody's help nor did any one extend it.

Shahram, a Tajik man from Parwan province, stated similarly: 'Asking help from those who are also passing through the same situation seems quite ridiculous. Nobody has enough time to think of others.'

Others felt entirely reliant on family or friends. While Kalsoom left her home in Farah province with her mother and sisters to a small village in Anar Dara, and then onto Kabul, displaced person Gul Sher, from Fayzabad travelled to Kabul, but this was worse than home. So he wanted to seek refuge among his relatives in Qandahar, but was trapped when he realised that he was now an enemy of a different cultural group who had aligned themselves with the United States forces in Qandahar. Gul Sher reflected on the importance of kinship support:

> We would have vanished long before had we not been supported by relatives and friends. In Afghanistan we have strong bonds of relationships and friendships. Many others like us were supported by relatives and friends.

Samiaullah from Nimroz province when narrating his experiences about taking refuge with relatives in a nearby village pointed out that being a Pakhtun-dominated area, the Taliban's visits were frequent. Subsequently, the local people began to suspect them. At first the family was considered to be friends of Russia and now loyal to the United States. As a result, he said, 'we had to leave that place also after a brief stay there and came to Nanghar in a very miserable condition'. Ahmad recalled the chaos in the following words:

> [T]he situation had worsened beyond imagination. It was before US attack on al Qaeda or Afghanistan. I am not sure but one of the groups either of Hib-e-Islami or Taliban was after us. We had a very huge business and it was really very difficult for us to leave all at once. Moreover, a big chunk of our money was still tied up in the market. But when life is in danger, money has little importance.

Most people perceived that their current location was only marginally safer than their own villages and towns. A common explanation was: 'Here in Kabul, the conditions are not very favourable, but comparatively better than ours back home.' It was inferred that this would be a temporary measure until returning home. While most did not aspire to emigrate, when considering this option, they preferred Uzbekistan, Kazakhstan, and Tijikistan as destinations as these countries were perceived to be culturally suitable and more liberal than neighbouring Pakistan or Iran. 'They are near to us. We are also familiar with their cultures', was a common saying. Additionally, Faiz, a Hazara, explained, 'in the past, we had migrated to these countries', while Shaihak, a Baluchi would select Uzbekistan as an option because 'they still respect us'. Omar, a Tajik added: 'I have been to Uzbekistan, Kazakhstan and Tajikistan. They are really very friendly and open hearted people. We do not consider ourselves strangers there.'

Our respondents tended not to mention the immediately neighbouring countries as potential places where they could migrate to because they felt they had handled the burden of various waves of migrations during different wars in the region and, economically, could not be expected to bear the burden without further support from the international community. For example, a Hazara Taxi driver stated:

> Giving shelter to someone for whole 30 years is indeed a difficult job. I have heard that Iran and Pakistan is already on the way to expel the migrants. It will be quite unfair to expect too much from these countries as they are already burdened by their own problems.

Moreover, neighbouring countries were becoming less of a migration option because, as some respondents with a cosmopolitan outlook explained, in the past, they had the opportunity to go to Central Asian States for higher education or employment. However, migrating to these states became progressively difficult due to certain activities of militants. Subsequently, the Afghanis were labelled as either extremists or terrorists. 'We face same kind of situation in Iran and Pakistan now,' noted Gulzaar. This is a broader problem too as many participants were afraid that the reception in these countries might not be very positive. According to a middle aged man,

> [w]e have very few friends or sympathizers in other countries. In the beginning, these countries accepted and received us with open hearts, but now things have changed. These countries are suspicious and a bit scared of us.

This unfair treatment was considered to be the underlying reason for many Afghani refugees in Pakistan returning home. In analysing their situation, the well-educated participants looked further abroad towards Western countries. Ahmad felt that he had reasonable chances of settlement abroad and would seek political asylum if appropriate business opportunities became available in France.

Suleman, a businessman would also select a Western country if given the option:

> My kids were used to a luxurious life and were insisting on shifting to America. Actually, the issue of my kids' education was compelling me to think over the option of taking refuge in some foreign country. We wanted to leave our area and move to US by hook or by crook because, I had heard that many of my friends who had left earlier than us were still struggling to survive in other areas [of Afghanistan].

However, even for the educated authorised migration as a refugee was not perceived to be easily achievable. Despite the difficulties in obtaining authorised migration, no Afghani interviewee was considering unauthorised travel. This stands in stark contrast to the sentiments of the Sri Lankan Tamils, which will be discussed later. According to the internally displaced persons (IDPs) in Afghanistan, the human traffickers required to facilitate this migration were perceived as untrustworthy. The following remark sums up the overall sentiments of respondents:

> [T]hey are fraudsters. They have cheated many people. Many people have made it to Europe, US, and Central Asian States by paying hefty amounts to these agents. But personally, I don't trust these agents. These agents are only good at hoodwinking helpless people like us.

Moreover, participants feared criminalisation. This was exemplified by drawing attention to the story of a family who hired a human trafficker who had promised to take them to Germany or Norway, but the agent disappeared in Peshawar. The Pakistani authorities arrested the family and sent them to jail to serve a sentence of seven months. Most people were aware of the consequences of paying people smugglers. It would lead to either arrest or loss of money, and, more seriously, they knew that once a person is arrested then he or she would be banned from that country. All respondents pointed to numerous cases of people being held in detention for many years instead of securing political asylum in Western countries.

Kalsoom Nazoo perceived that a peaceful and orderly life would not be achieved by breaking rules to reach that safer destination: 'I will respect their law because I myself am fed up with lawlessness here. I can only travel with proper documents.' Clearly, for many people the financial resources required were beyond their reach. One informant explained:

> I am the head of nine family members and they ask about 5 million for the whole family, which is beyond my affordability. Secondly, these agents do not give you any guarantee whether I will get settled abroad or not. This is only possible when it is done on a state level. And I see no hope of that.

Although some respondents felt that theoretically the only secure option was the internationally coordinated, legal, migration, they were thoroughly sceptical of it coming to fruition. This issue will be taken up in greater detail later in this chapter. Our findings from Afghanistan challenge the conventional understanding of authorised migration coordinated by the international community as the safest and best option, since participants repeatedly emphasised that they were not hopeful that the international community would step in to aid their safety. Firoza, a teacher lamented:

> The international community turned a blind eye to the problems of peaceful Afghans like us. They left us at the mercy of these ignorant people. We had many hopes from the international community but they were only after their vested interests due to which our country is at the verge of destruction.

In some cases people felt that international intervention had destabilised Afghanistan in the first place, leading to the continual conflict and the displacement they experience today. This was a strongly held view with comments, such as, 'Foreign interference is the root of all evil' and 'If foreign interference is ended today, peace will return to this unfortunate land in days not in months'. Nousafarin was emphatic: 'We were living happily and peacefully with one another for centuries but when foreign powers started interfering in the situation deteriorated beyond imagination.' The respondent added that peace and safety in Afghanistan would only be achieved if 'the people of Afghanistan realize that we have to build this country on our own and not with the help of some foreign power which has its own vested interests'. Haamein, on the other hand, felt that the international community could facilitate peace, but with appropriate solutions, not merely sending aid: 'I ask one favour from international community to please bring peace to our country, instead of useless funds.'

Against popular misconceptions about motivations of asylum seekers, it is interesting to note that some Afghanis did not want to go anywhere; rather they had the overwhelming desire for conditions to return to a state of normalcy and security. 'Migrations are no solution. I ask one thing from the powers that be.... I appeal to them to come and work sincerely for the rebuilding of our beloved country,' implored Kalsoom. Some were passionate about their homeland and said: 'I will never leave Afghanistan. It is my mother country and very beautiful'. Golzaar reflected on the futile nature of migrating from place to place: 'Sometimes, it seems as if our whole life would be wasted in shifting from one place to another.' Ghulam summed it up succinctly when he said 'it is indeed very disheartening to leave one's own land'. Yet, they all recognised that it was still too violent to return home.

Overall, the internally displaced Afghanis in our study felt helpless about their personal circumstances and did not demonstrate a sense of empowerment regarding their future. Ideally, they hoped for a return to home if peace prevailed, and felt that finding peace in a foreign country was an unattainable goal not worth serious consideration—particularly when there was little logistical support on the ground. Their focus was on day-to-day survival. For those for whom migration is to be an option, they would like it to be safe and have the desire to return home once stability is achieved. These accounts also dispel the popular myth prevailing in Western countries that asylum seekers are inherently economic migrants seeking a better life style.

Sri Lankan Experiences of Displacement

Unlike in Afghanistan, where many respondents attributed their troubles to be entirely due to the outcome of several decades of international intervention, Sri Lankan Tamils felt intensely insecure as an identifiable ethnic minority, with the civil war exacerbating their misery. The ongoing police brutality is a key factor in their desire to flee elsewhere. However, many, who had past associations with Tamil resistance groups, were quick to point to the current insecurity stemming from the factions emerging within these organisations, while others, who had no specific allegiances to any of them, held them partly responsible for the threats against them. Yet, they too maintained that as a minority they were under constant surveillance and faced repeated assaults by militias, the army, and security forces.

Since Independence, as a minority ethnic group, Sri Lankan Tamils have been progressively marginalised. Some of our respondents as idealistic young persons had joined either the Tamil Eelam Liberation Organization (TELO) or the Liberation Tigers of Tamil Eelam (LTTE). However, since the end of the Eelam War in May 2009, many factions have emerged; one TELO movement member has become a Member of Parliament (MP), and some former LTTE members now assist the government to identify other ex-LTTE cadres and ex-members of the paramilitary groups. The following vignette highlights complexities of the threats against families.

Murugan is a former member of the TELO paramilitary group, but now works as an organiser for the MP Vinogaralingam. Previously, Murugan had fled to Jeddah temporarily, leaving his family in Mannar district in the northern province of Sri Lanka. There they received threats that if Murugan was not returned to the TELO, the whole family would be kidnapped or killed. In 2010, when Murugan was campaigning on behalf of an MP, he started getting threats. By then the TELO movement had split into two groups and the new TELO began to pressure Murugan (as an old TELO) member to join them. His refusal resulted in threats and assaults:

> In December I had gone to Trincomalee Town on the motorbike to purchase some rations for the house. On my return I was stopped by a group of strangers and they assaulted me with wooden poles. When I screamed in pain, the strangers just ran away and vanished. I returned home and underwent native treatment for the shoulder dislocation, but remained silent as I was afraid to go to the hospital or make a complaint to the Police. The threats have continued … at first I was not sure of the reasons, but now I realize that we are being threatened because I was from the TELO paramilitary group. But the threats are not just from one party or group; possibly it could be the new faction of the TELO and on the other side it could be the government forces. There is a high level of insecurity for the family and in this situation we are forced to leave our home and hide from our enemies as we are not even able to go before the law or seek legal measures.

The cases of Sekharan and Henrietta also illustrate the continuing uncertainties and threats faced by those caught up in the civil war. In 2006, as teenagers, Sekharan and Henrietta were forcibly abducted by LTTE cadres in Triconamalee and Jaffna, respectively. While Sekharan escaped shortly afterwards and lived in the Utchampatti Welfare Centre in India, Henrietta managed to run away after two years, returning home and surrendering herself to the Tellipalai Rehabilitation Centre in Jaffna, managed by the Government Military Forces. In the case of Sekharan, he returned to Sri Lanka along with several families in late 2010 with the support of the United Nations High Commission for Refugees (UNHCR). However, his mother was unable to find him at Colombo airport. When she made inquiries at the UNHCR sub-office in Triconamalee, she came to learn that he was arrested by the Terrorist Investigation Division. It transpired that UNHCR were not informed of his arrest. None of the UNHCR officials had been present at the airport to check the names on the list of the persons arriving that day. While Sekharan's mother regrets not sending her son abroad via illegal means, Henrietta's plight highlights the aftermath of a war that has 'officially' ended, but continues to haunt many people attempting to re- adjust. As Henrietta explained:

> I want to live a normal life that a girl would usually want to live, but most people do not want to accept me as a normal girl and because of this my attempt to find a life partner too has failed. Even though there had been marriage proposals from Tamil boys living in the West, these proposals always fall through once they

> find out that I am an ex-combatant. According to the Tamil culture, I have to get married soon so that my sisters too could be given in marriage at the right age. My parents are also unhappy due to this and fear that I may have to face threats in life if the situation in the country changes again. I don't want to leave the country, but my parents are pressurising me to go to the West so that the stigma will disappear. It will automatically erase the image that others have about me.

Although her family cannot afford to send her abroad legally, she is planning to migrate illegally to the West. She is unaware of the procedure to travel illegally, but has made up her mind to go. In her view, since there are many North and East Sri Lankan Tamils living in Australia, Canada, and the UK, she feels she should migrate to any of these countries soon in order to fulfil her parents' wish and also find herself a new life.

Afghani and Sri Lankan Tamil Response to UN 'Benevolence' and 'Rescue'

The disdain towards the role of international organisations, particularly the UN bodies, came into sharp relief when we raised the issues of assistance in resettlement. The UN, as an impartial entity to protect the well-being of peaceful Afghanis caught up in the conflict, was not viewed as particularly effective. It was perceived as apathy on behalf of the international community. While Gul Sher simply did not know of or understand their function, Suleman recognised the logistical challenges facing aid agencies. Likewise, Kalsoom Nazoo understood the problems facing the UN. When she did try for help, it was her persistent efforts that 'made it possible for us to procure some edible items from UNHCR'. She was the only person to report a positive encounter. Samiaullah remarked with deep sarcasm whether it was even appropriate for the UN to provide assistance:

> In my opinion, it would be asking too much from UN to bring peace to our country. Sometimes, I think that these organizations have made it their business to extract money from the donors in the name of sending people abroad.

Indeed, a number of people inferred that international organisations on the ground in Afghanistan were corruptible and help was more attainable for people who had political connections and some power in the situation. Faiz felt that '… all their help is availed by leaders, warlords, commanders and many other influential people. In the end, tthere is little left of relief goods for poor people like us.' In terms of opportunities to migrate, Omar believed that people who had migrated to foreign countries 'had political connection while we had none'.

Refugee services were perceived by all participants as unattainable with systems and processes making it impossible to meet the criteria for migration as a refugee. Most were not familiar with the processes; the following comment by a woman in her early 30s typifies the perceptions of respondents:

> [T]here are many loopholes in their procedures; otherwise people would not face such kind of hardships. The IDPs in Kabul and many others living in the camps on the other side of the border complain about lack of proper arrangements on the part of the international organizations … migrants in Pakistani camps are in a miserable condition. Whereas, the common perception is that migrants on the Iranian side are in a comparatively better condition.

According to the overwhelming majority of the respondents, it was impossible for lay people 'like us' to meet the strict criteria set by UN missions. Ahmed has observed and has heard that migration is near impossible even after the tiring and lengthy procedures that are compulsory. In his view, '[i]f any UN mission was serious or active in this regard we would have been settled long ago'. As a medical practitioner, Meherzad has registered to migrate; but he noted with exasperation the following:

> I have filed my papers three times, but am still on waiting list. I have been called twice for interview. I think something is lacking in my case to qualify their requirement. If an educated person like me is passing through such ordeal then you can imagine the fate of uneducated ones.

Most participants were unfamiliar with the processes, except to understand that it was complicated and those who had applied experienced it as an 'ordeal'. Banki (2008: 9; see also Doornbos et al. 2001) writes of the confusion and anxiety that can affect the capacity of refugees to apply for asylum through regular channels.

> There is also confusion surrounding the process of applying and interviewing for resettlement. Depending on the resettlement country, refugees may be asked to meet with different personnel (UNHCR, resettlement country representatives and IOM staff) as many as five times. The process, from first interview to flight departure, is expected to take six months on average, if no extenuating circumstances present themselves. But some refugees confuse an expression of interest (made by letter or in person at UNHCR) with an interview, and the wait thereafter feeds their anxiety.

Many of our respondents in Sri Lanka had made repeated attempts to seek the help of the International Committee of the Red Cross (ICRC), UNHCR, and the Human Rights Commission of Sri Lanka office in their respective areas requesting these institutions to offer a letter to them in order to apply for asylum in a country like Canada or Australia. Ultimately, however, these institutions had refused their requests and informed them that they were not in a position to submit such a letter to these particular individuals. For example, earlier this year, Muralidaran had been blindfolded and kidnapped by vigilante groups. In his absence, his wife and children were assaulted. After fleeing his captors, he sought the assistance of the ICRC, but was advised that ICRC could not investigate this incidence given that the Eelam War had now come to an end. He continues to receive threats from 'unknown persons'. Thus, such participants in our study could not understand why their request for asylum was refused, 'whereas some others who are more secure than us have got selected to go to different countries without much difficulties,' queried Devika, a 35-year-old volunteer health care worker and a mother of three children.

A number of Sri Lankan people had also tried to migrate to the UK and Canada, but their applications were not successful. In some instances they did not receive the sponsorship from relatives they had hoped to secure. In other instances, their bank balances did not support the requirements of the UK High Commission. Although most had relatives in India and were being constantly asked to join them there, in general, Sri Lankan Tamils did not see this as a viable option for their children's future. A frequently used explanation was:

> Already in India there are Sri Lankan Tamils living as refugees, but this is not what I want for my children. If an opportunity comes to go towards the West, we would do so as this is a more secure environment than in Sri Lanka or India.

In this regard, their perspectives resembled those of the Afghan respondents who argued that neighbouring countries were suffering from 'compassion fatigue' and that there was little future for refugees. However, unlike the Afghanis, the Sri Lankan families were deeply upset at their failure to secure asylum despite repeated earnest attempts; it led them to conclude that they would resort to illegal means given that legal pathways failed. The case of Chinnappan, a farmer in his early 40s, typifies this dilemma. Although he had never had any political affiliations, Chinnapan was abducted and severely beaten by strangers. His family members were continually threatened. He made several attempts to seek asylum. He had previously worked legally in Qatar in the late 1990s. However, he noted, 'because of my age I can't get a work permit.... I'm planning to travel illegally to Italy and then will try to proceed to [another] Western country.' It seems that the protracted processes of asylum seeking among our Sri Lankan participants led to resentment and despair. According to Jansen (2008), seeking to improve the quality of life does not mean that the displaced are not desperate. Poverty is itself a form of insecurity, and, as Battersby (2008: 16) remarks, it is hardly surprising that 'people with sufficient economic means will take extraordinary steps to ensure that they and their families find safe haven'.

Last Resort: Unauthorised Migration

When facing peril to themselves and those closest to them, people must take action, and such action often necessitates making a choice to engage with conditions of high risk (Hayenhjelm 2006). Robinson and Segrott (2002) note the need for awareness that the term 'choice' is contentious when used in the context of asylum seeker migration. For asylum seekers, they emphasise that

> … personal decision making is rarely a rational exercise in which people have full knowledge of all the alternatives and weigh them in some conscious process designed to maximise returns.… Instead, [they are] active agents who search out both information and contacts and change, circumvent, and create institutions in order to achieve desired objectives. (Robinson and Segrott 2002: 6–7)

Undocumented international migration is a hazardous undertaking, with the risks and costs of travel increasing with the level of 'clandestinity' (Düvell 2008: 492). Precluded from using regular and safe means of transport, undocumented migrants are at risk of drowning at sea, freezing to death in aircraft undercarriages, asphyxiating in lorries, being killed on roads as they leave trucks, or being electrocuted or falling as they cling to the roofs and sides of trains (Athwal and Bourne 2007). Taking to the oceans in cramped, often unseaworthy, vessels is a patently hazardous activity, and those attempting it have a low chance of surviving the experience unscathed, if at all. Compounding the risk of storm damage, drowning, and starvation is the threat of attacks by pirates, who are known to rob, rape, and/or murder those without protection on the sea (Kirkpatrick 1988: 400). Asylum seekers are commonly at the mercy of unscrupulous people smugglers, who transport them under inhumane conditions. They can be crammed into inadequate spaces or locked in freight containers without sufficient air, water, or food, and may be abandoned en route if detection by authorities appears imminent (Schloenhardt 2003: 139).

Despite these hazardous journeys, our Sri Lankan informants were prepared to put themselves in great danger in their search for freedom from war and persecution. They emphasised that unauthorised travel relying on the use of people smugglers (termed 'agents' by Robinson and Segrott 2002) was one of the options that was relatively freely available to them. Although most had limited financial means, they were prepared to borrow, sell their possessions, or negotiate with the agents to pay a portion later when they reached safety. According to Loescher and Milner (2003), the services of smugglers become more valuable when receiving countries limit access to state protection and opportunities, and people smuggling is consequently on the rise around the world (Antonopoulos and Winterdyk 2006). We have already demonstrated the difficulties in accessing official channels and the respondents' perceptions of their ineffective approaches. In contrast, human smugglers are often highly visible and easy to contact directly or through social networks (Koser and Pinkerton 2002; Doornbos et al. 2001). Specific services performed by agents include providing travel documents, facilitating journeys (sometimes accompanying refugees and reclaiming false documentation at the destination point) and channelling individuals towards or away from particular countries (Robinson and Segrott 2002).

Robinson and Segrott's (2002) study showed that the vast majority of asylum seekers assisted by 'agents' had originated from Sri Lanka and Iran. These respondents reported that, without the agents, they would have had no hope of being able to escape their country of origin and reach a place where they could claim asylum. Although our research found no Afghanis willing to undertake irregular migration, it is worth dwelling on the situation of Afghan asylum seekers embracing this path. Cashmore (2006) recounts the story of a young Afghani man whose only chance of escaping abduction by the Taliban had been with a people smuggler known to a contact of his father. From such powerless positions, many asylum seekers are compelled to cede control over logistical decision making, with some having no idea where they were going until after they had arrived. Some agents—especially in Sri Lanka—had previously negotiated with older relatives of asylum seekers, but the asylum seekers themselves had not been consulted about potential destinations. Others had more negotiating power, depending on their own capacity to pay and the agent's capacity to deliver favoured destinations. In some instances, only one destination was on offer by smugglers and asylum seekers needed to seize that opportunity or risk further persecution or deportation if they remained (Robinson and Segrott 2002).

For those seeking to travel on the open sea, smugglers will usually be their only option. Hoffman's (2007) research documenting asylum seekers travelling in boats arranged by smugglers points to a disjuncture in accounts given by academic researchers and the Australian Federal Police regarding the scale and nature of people smuggling in the Asia Pacific region. While academic sources report a prevalence of loose, informal networks based on ethnic or kinship ties, police sources depict smuggling syndicates as criminal gangs. Common language and ethnicity facilitates pragmatic communication between client and agent, Hoffman argues, but it does not necessarily lead to greater trust. Smugglers are overwhelmingly driven by the prospect of economic gain and migrants with few monetary resources and often taken for short distances or over more dangerous routes (van Liempt and Doomernik 2006). Smuggled migrants may arrive at destination countries with no money after paying the agents (Koser and Pinkerton 2002), placing themselves at even higher risk. While some former asylum seekers attribute humanitarian motives to the agents they used (Doornbos et al. 2001; Neske and Doomernik 2006; van Liempt and Doomernik 2006), in the majority of cases, as we have highlighted, they are utilised, not because they have any particular personal qualities, but because no other effective alternatives were available. As Hayenhjelm (2006: 192–3) observes,

> … risks from vulnerability need social and political attention rather than good advice to individuals, since it is the very lack of alternatives that makes people take these risks. If these conditions are not altered the risks will be taken nonetheless.

Nevertheless, in light of the recent tragedy when a boat carrying refugees capsized off Ashmore Reef in northern Australia and the resulting deaths, the Human Rights Commissioner in Sri Lanka who had assisted us to carry out the interviews noted that he wanted to start information campaigns to warn potential asylum seekers of the perils of such journeys. Rather than exercising a judgemental approach towards undocumented migration, his main concern was the exploitation of vulnerable people by smugglers. Whether such an approach would have deterred any our respondents is rather debatable.

The case of Perumal illustrates the tragic consequences of seeking the assistance of human traffickers and the ensuing failed asylum claim. Permual's family fled their home in the 1990s to Chilaw in the North western province. During this period, to support his family, he had made several attempts to secure overseas contract work, but failed. When they returned to their home village in Jaffna in the mid-2000s, he not only began receiving threats, but was also arrested despite having no political affiliations to Tamil secessionist groups. His arrest meant that he could not obtain a police clearance for legal overseas travel and, therefore, sought the assistance of people smugglers to go to Britain. Unfortunately, his claim for asylum did not eventuate and he ended up in detention and was deported.

As a labourer with limited means, Perumal lamented about what he perceived to be the arbitrariness of those being granted asylum.

> I have now lost all my money and also the belongings that I possessed before I left to UK. People I met in London have also paid well known smugglers and got false clearance documents prepared in Sri Lanka in order to obtain a visa. After going there these people seek Refugee Status and it is also granted to most of these people—except for a very few. There is no proper inquiry made about the people seeking Refugee Status. The authorities in UK do not follow any proper procedure to grant or reject the Refugee Status of these people. I feel that they just pick and choose as per their interest. It is also luck that plays a role in here. Now as I have been deported, I will never be able to travel again to any country outside Sri Lanka. This is a real difficult situation for me and my family. I have the responsibility of taking care of both my mother who is a widow, and as well as my wife's mother who is a disabled person. It is difficult to find jobs in Jaffna. I do not know how to survive after this heavy loss. I have lost my reputation too.

Had Perumal been aware of the intricacies of people smuggling and the limited possibilities of securing asylum, it is difficult to imagine whether the information would have acted as a deterrent.

Conclusion

Popular and political discourses continue to routinely condemn asylum seekers for not adhering to rules of law or of 'fair play'. However, as we have shown throughout this chapter, the presence of a neat and methodical structure for decision-making implied in this notion does not apply to the chaotic conditions experienced by many asylum seekers. Battersby (2008: 15) sums up this argument succinctly:

> Both the policies and the rhetoric deployed by Western governments against asylum seekers are designed to dampen public sympathy for genuine and extreme human suffering. The chaotic nature of refugee dispersal and ... the resettlement process does not function according to developed country notions of efficient social service.

Our findings in Sri Lanka and Afghanistan affirm a number of the arguments put forward by several researchers. First, as the former Liberal Party MP Petro Georgiou (cited in Gordon, 2005: 9) suggested, '[i]f the uninvited offend against our preference for an orderly migration process', attending to the stories of asylum seekers themselves can 'persuasively elucidate why escaping from persecution is not an orderly process'. Second, clearly the spaces occupied by forced migrants are 'typically defined by social chaos ... where affected populations experience a profound sense of confusion and disorientation' (Rodgers 2004: 48). Third, Silove's (2002) contention that not only are the opportunities to access sites of regularised authority limited, the chances of a positive outcome are also slight. Our research also dovetails with the findings from the Edmund Rice Centre for Justice and Community Education (ERC) (2009: 1), that highlights the unfortunate reality that the imagined orderly world where asylum seekers are protected elsewhere until they are selected and settled in Australia is by no means a certainty.

In the final analysis, the value of investigating the circumstances of displaced people in their countries of origin from the two dominant groups currently seeking asylum in Australia, the Afghanis and the Sri Lankan Tamils provide us the ability to better understand the differing contexts for their internal displacement and the factors that make displaced people candidates for seeking asylum. It offers us an insight into the vulnerability of specific communities towards human trafficking as a mechanism to achieve this. This approach has facilitated the comparison of differences and similarities between these two groups which, once in Australia, are often perceived by Australian society as one, homogeneous, group known as 'boat people'. By understanding their varied responses and decision-making processes while still in country, key learnings could then emerge to inform this complex issue and its management.

References

AAP General News Wire. 2009. More Sri Lankan asylum seekers expected. *The Age*, Melbourne, AAP General News Wire, 1.

Agier, Michel. 2002. Still stuck between war and city: A response to Bauman and Malkki. *Ethnography*, 3(3): 361–6.

Antonopoulos, Georgios A. and Winterdyk, John. 2006. The smuggling of migrants in Greece: An examination of its social organization. *European Journal of Criminology*, 3(4): 439–61.

Athwal, Harmit and Bourne, Jenny. 2007. Driven to despair: Asylum deaths in the UK. *Race and Class*, 48(4): 106–14.

Banki, Susan. 2008. *Bhutanese Refugees in Nepal: Anticipating the Impact of Resettlement*, A Place to Call Home: Briefing Paper [Online]. Available at: http://www.austcare.org.au/media/56970/arcnepalbp-lowres.pdf [accessed 20 October 2010].

Barsky, Robert F. 1995. Arguing the American dream à la Canada: Former Soviet citizens' justification for their choice of host country. *Journal of Refugee Studies,* 8(2), 125–41.

Battersby, Paul. 2008. A world turned upside down: Risk, refugees and global security in the 21st century, in *Asylum Seekers: International Perspectives on Interdiction and Deterrence*, edited by Alperhan Babacan and Linda Briskman. Newcastle, UK: Cambridge Scholars Publishing, pp. 10–27.

Betts, Alexander. 2003. *The Political Economy of Extra-territorial Processing: Separating 'Purchaser' from 'Provider' in Asylum Policy*, New issues in refugee research Working Paper No. 91. Available at: http://www.unhcr.org/3efc0ea74.pdf [accessed: 24 November 2010].

Bleasdale, Lydia. 2008. *Under Attack: The Metaphoric Threat of Asylum Seekers in Public-political Discourses*. Available at: http://webjcli.ncl.ac.uk/2008/issue1/bleasdale1.html [accessed: 22 November 2010].

Blommaert, Jan. 2009. Language, asylum, and the national order [with comments]. *Current Anthropology*, 50(4): 415–41.

Briskman, Linda and Cemlyn, Sarah. 2005. Reclaiming humanity for asylum-seekers: A social work response. *International Social Work*, 48(6), 714–24.

Brown, Nik and Michael, Mike. 2003. A sociology of expectations: Retrospecting prospects and prospecting retrospects. *Technology Analysis & Strategic Management*, 15(1), 3–18.

Cashmore, Judy. 2006. Yasser, in *Fenced Out Fenced In*, edited by Natalie Bolzan, Michael Darcy, and Jan Mason. Altona: Common Ground Press, pp. 9–12.

Castles, Stephen. 2007. The factors that make and unmake migration policies, in *Rethinking Migration: New Theoretical and Empirical Perspectives*, edited by Alejandro Portes and Josh DeWind. New York: Berghahn Books, pp. 29–61.

Clarkson, Petruska. 2003. War; bystanding and hate—Why category errors are dangerous. *Psychotherapy and Politics International*, 1(2): 117–32.

Collyer, Michael. 2010. Stranded migrants and the fragmented journey. *Journal of Refugee Studies*, 23(3): 276–93.

Crawley, Heaven. 2006. Forced migration and the politics of asylum: The missing pieces of the international migration puzzle? *International Migration*, 44(1): 21–26.

de Haas, Hein. 2007. Turning the Tide? Why Development Will Not Stop Migration *Development and Change* 38(5): 819–41.

De Giorgi, Alessandro. 2010. Immigration control, post-Fordism, and less eligibility: A materialist critique of the criminalization of immigration across Europe. *Punishment and Society*, 12(2): 147–67.

Doornbos, Nienke, Kuijpers, Anne Marie, and Shalmashi, Khalil. 2001. *Refugees on Their Way to a Safe Country*. Available at: http://home.medewerker.uva.nl/n.doornbos/bestanden/Refugees%20on%20their%20way%20to%20a%20safe%20country.pdf [accessed: 19 November 2010].

Düvell, Frank. 2008. Clandestine migration in Europe. *Social Science Information*, 47(4): 479–97.

Dwyer, Peter. 2008. *Meeting Basic Needs? Exploring the Survival Strategies of Forced Migrants*. Available at: http://www.liv.ac.uk/law/rdi/docs/Dwyer_WS1_Material_ORIGINAL.pdf [accessed: 23 November 2010].

Edmund Rice Centre for Justice and Community Education (ERC) and The Australian Catholic University. 2009. Shifting the Focus from people smugglers to protection. *Just comment*, 12(3):1-2.

Fairclough, Norman. 2003. 'Political correctness': The politics of culture and language. *Discourse and Society*, 14(1): 17–28.

Fassin, Didier and d'Halluin, Estelle. 2007. Critical evidence: The politics of trauma in French asylum policies. *Ethos*, 35(3): 300–29.

Fekete, Liz and Webber, Frances. 2010. Foreign nationals, enemy penology and the criminal justice system. *Race and Class*, 51(4), 1–25.

Ganguly-Scrase, R, Vogl, G and Julian J. 2006. Neoliberal Globalisation and Women's Experiences of Forced migrations in Asia. *Refugee Watch: A South Asian Journal of Forced Migration,* 28: 1–18.

Gordon, M. 2005. *Freeing Ali: the human face of the Pacific solution*, University of New South Wales Press, Sydney.

Innes, A. J. 2010. When the Threatened Become the Threat: The Construction of Asylum Seekers in British Media Narratives. *International Relations,* 24(4): 456–77.

Hayenhjelm, Madeleine. 2006. Out of the ashes: Hope and vulnerability as explanatory factors in individual risk taking. *Journal of Risk Research*, 9(3): 189–204.

Hörnqvist, Magnus. 2004. The birth of public order policy. *Race and Class*, 46(1): 30–52.

Hoffman, Sue. 2007. *Reaching Australia: Iraqi asylum seekers in transit in South-East Asia*. [Online: Second Multi-disciplinary Conference of International Association of Contemporary Iraqi Studies (IACIS) Philadelphia University, Amman, Jordan]. Available at: http://www.safecom.org.au/hoffman2007.htm [accessed: 22 October 2010].

Jansen, Bram J. 2008. Between vulnerability and assertiveness: Negotiating resettlement in Kakuma refugee camp, Kenya. *African Affairs*, 107 429: 569–87.

Kirkpatrick, J 1988 *Legitimacy and Force: National and international dimensions,* Transaction Books, New Brunswick, New Jersey.

Koser, Khalid and Pinkerton, Charles. 2002. The social networks of asylum seekers and the dissemination of information about countries of asylum

[Online]. Available at: http://rds.homeoffice.gov.uk/rds/pdfs2/r165.pdf [accessed: 25 November 2010].

Kunz, E.F. 1973. The refugee in flight: Kinetic models and forms of displacement. *International Migration Review*, 7(2): 125–46.

Limbu, Bishupal. 2009. Illegible humanity: The refugee, human rights, and the question of representation. *Journal of Refugee Studies*, 22(3): 257–82.

Linden, Sonja. 2007. Manifesto: Reclaim asylum. *Index on Censorship*, 36(2): 123.

Lindley, Anna. 2010. Leaving Mogadishu: Towards a sociology of conflict-related mobility. *Journal of Refugee Studies*, 23(1): 2–22.

Loescher, Gil and Milner, James. 2003. The missing link: The need for comprehensive engagement in regions of refugee origin. *International Affairs*, 79(3): 583–617.

Malkki, Liisa H. 1995. *Purity and Exile: Violence, Memory, and National Cosmology among Hutu Refugees in Tanzania*. Chicago: University of Chicago Press.

McKee, Alan. 2003. *Textual Analysis*. London: Sage Publications.

Moore, Will H. and Shellman, Stephen M. 2007. Whither will they go? A global study of refugees' destinations, 1965–1995. *International Studies Quarterly*, 51(4): 811–34.

Neske, Matthias and Doomernik, Jeroen. 2006. Comparing notes: Perspectives on human smuggling in Austria, Germany, Italy, and the Netherlands. *International Migration*, 44(4): 39–58.

Pearce, Julia M. and Stockdale, Janet. 2009. U.K. responses to the asylum issue: A comparison of lay and expert views. *Journal of Community and Applied Psychology*, 19(2): 142–55.

Pedersen, Anne, Watt, Susan, and Hansen, Susan. 2006. The role of false beliefs in the community's and the federal government's attitudes toward Australian asylum seekers. *Australian Journal of Social Issues*, 41(1): 106–19.

Porter, Elisabeth. 2003. Security and Inclusiveness: Protecting Australia's Way of Life, *Peace. Conflict and Development*: 3, July, 1-18.

Poynting, Scott and Mason, Victoria. 2007. The resistible rise of Islamophobia: Anti-Muslim racism in the UK and Australia before 11 September 2001. *Journal of Sociology*, 43(1): 61–86.

Rees, Susan and Silove, Derrick. 2006. Rights and advocacy in research with East Timorese asylum seekers in Australia. *Journal of Immigrant and Refugee Studies*, 4(2): 49–68.

Robinson, Vaughan and Segrott, Jeremy. 2002. *Understanding the decision-making of asylum seekers*, Home Office Research Study 243 [Online]. Available at: http://rds.homeoffice.gov.uk/rds/pdfs2/hors243.pdf [accessed: 23 November 2010].

Rodgers, Graeme. 2004. '*Hanging out' with forced migrants*: Methodological and ethical challenges. *Forced Migration Review*, 21, September, 48–49.

Rose, Peter I. 1981. Some thoughts about refugees and the descendants of Theseus. *International Migration Review*, 15(1/2): 8–15.

Sabat, S.R. 2003. Malignant positioning and the predicament of the person with Alzheimer's disease, in *The Self and Others: Positioning Individuals and Groups in Personal, Political, and Cultural Contexts*, edited by F.M. Moghaddam and R. Harré. Westport, CT: Greenwood Publishing Group, pp. 85–98.

Sales, R. 2002. The deserving and the undeserving? Refugees, asylum seekers and welfare in Britain. *Critical Social Policy* 22(3): 456-78.

Schloenhardt, Andreas. 2003. *Migrant Smuggling: Illegal Migration and Organised Crime in Australia and the Asia Pacific Region*. Boston, MS: Martinus Nijhoff.

Shawcross, William. 1979. Refugees and rhetoric. *Foreign Policy,* 36, Fall: 3–11.

Sheridan, Lynnaire M. 2009. *'I Know It's Dangerous': Why Mexicans Risk Their Lives to Cross the Border*. Tucson: University of Arizona Press.

Silove, Derrick. 2002. The asylum debacle in Australia: A challenge for psychiatry. *Australian and New Zealand Journal of Psychiatry*, 36(3): 290–96.

Sinapi, Michèle. 2008. The displacements of the `shadow line'. *Social Science Information*, 47(4): 529–39.

Stalker, Peter. 2001. *The No-nonsense Guide to International Migration*. Oxford: New Internationalist Publications.

Strauss, A. and Corbin, J. 1994. Grounded theory methodology: An overview, in *Handbook of Qualitative Research*, edited by N. Denzin and Y. Lincoln. Thousand Oaks: SAGE Publications, pp.1-18

Taylor, S. and Bogdan, R. 1998. *Introduction to Qualitative Research Methods: A Guidebook and Resource*. New York: John Wiley & Sons.

Tazreiter, Claudia. 2008. Asylum. *Encyclopedia of Race, Ethnicity, and Society* [Online]. Available at: http://www.sage-ereference.com/ethnicity/Article_n44.html [accessed: 6 May 2010].

Turner, S. 2004. Under the gaze of the 'big nations': Refugees, rumours and the international community in Tanzania. *African Affairs*, 103(411): 227–47.

Turner, Brian. 2010. Enclosures, enclaves, and entrapment. *Sociological Inquiry*, 80(2): 241–60.

van Liempt, Ilse and Doomernik, Jeroen. 2006. Migrant's agency in the smuggling process: The perspectives of smuggled migrants in the Netherlands. *International Migration*, 44(4): 166–90.

Welch, Michael and Schuster, Liza. 2005. Detention of asylum seekers in the US, UK, France, Germany, and Italy: A critical view of the globalizing culture of control. *Criminal Justice*, 5(4): 331–55.

Wodak, Ruth, Baker, Paul, Gabrielatos, Costas, KhosraviNik, Majid, Krzyzanowski, Michal, and McEnery, Tony. 2008. A useful methodological synergy? Combining critical discourse analysis and corpus linguistics to examine discourses of refugees and asylum seekers in the UK press. *Discourse and Society*, 19(3): 273–306.

Zimmermann, Susan. 2010. *Why seek asylum? The roles of integration and financial support. International Migration*, 48(1): 199–231.

Index